AN INTRODUCTION TO

Exploration Economics

Second Edition

AN INTRODUCTION TO

Exploration Economics

Second Edition

R. E. Megill

The
Petroleum Publishing Company
Tulsa, Oklahoma 74101

Books by Robert E. Megill.

How to Be a Productive Employee, 1973

An Introduction to Risk Analysis, 1977

Published by
Petroleum Publishing Company
Tulsa, Oklahoma 74101

Library of Congress No. 7870549
ISBN 0-87814-096-4
Printed in the U.S.A.

1 2 3 4 5 83 82 81 80 79

This book is dedicated to the two men who
guided my early search in the economics of
petroleum exploration—
J. R. Arrington and G. D. Priestman
and to the
Exxon Company, U.S.A.,
under whose auspices the search was made.

Contents

List of Figures

Preface

Exploration economics does not exist as a separate branch of either academic or practical economics.

It is a strange mixture of engineering economics, mathematics and statistics, probability theory and the more normal sciences of exploration—geology and geophysics. There are textbooks which cover each aspect. Nowhere, however, can one find a reasonable starting point encompassing the economics of the search for hydrocarbons. This book is a starting point. It is not a scholarly text on economics; it is a resumé of practical evaluating knowledge written especially for the working explorationist and the student geologist.

The book is written for the explorationist who knows little about the subject and who has forgotten most of his college math. It is intended to be truly introductory. As such the accomplished investment analyst will find this work elementary. He will find certain illustrations which oversimplify or generalize. However, some oversimplification must occur in a primer. Each generalization is footnoted with proper cautions if needed. None of the simplifications suggested will hinder sound investment evaluation.

It is my hope that the interested reader will continue his search amid available literature. The bibliography included at the end of each chapter should assist in that search.

Preface

As it became evident that a fourth printing of "An Introduction to Exploration Economics" was necessary, the author and publisher considered the advisability of a new version. This revised edition is the result.

The basic additions and corrections which make up the revised edition are listed below:

1. The increasingly inhibiting role of regulation is reviewed primarily because of its effect on delay. In the past decade some regulatory delays have been so long as to lower profits substantially, raise costs, or both.

   Anyone who thinks the logic in the revised edition overstates the conditions afflicting the industry should read William E. Simon's book, "A Time For Truth."[1] By comparison the author's comments are conservative relative to a person who in Washington, D.C., saw the real enormity of the problems generated by regulation.
2. The chapter entitled "Understanding the Terms" was revised to include the new standards set forth by Statement No. 19 of the Financial Accounting Standards Board.
3. The latest tax concepts are incorporated including a de-emphasis of percentage depletion. Although percentage depletion still has meaning for small companies, the problem in the book, associated with the School prospect, has been reworked to exclude percentage (but not cost) depletion.

4. Appropriate changes to accommodate recent developments have been made to the chapter on cash flow analysis and present value concepts.
5. A new addition on evolving yardsticks has been added to Chapter 7. Charts and curves are updated as needed.
6. A companion volume ("An Introduction to Risk Analysis") which is essentially an expansion of Chapters 8 and 9 was published in 1977.[2] Nevertheless, changes were made in both chapters on risk analysis as needed.
7. The most significant expansion was made to Chapter 10. Sections on inflation, sampling without replacement, methods on sensitivity testing and other additions make this chapter much more useful.

In general the revised edition includes sections and additions which respond to the questions most often asked in teaching the subject—both within the author's company environment and in industry.

However, no one can write a textbook involving tax laws and government regulations and expect it to remain up-to-date for long. The revised edition brings the reader the changes through 1978. With the vast increase in regulations streaming forth from both the state and Federal levels, only history will determine what unforeseen changes will affect the accuracy of the revised edition.

BIBLIOGRAPHY

1. Simon, William E., "A Time For Truth," McGraw-Hill Book Co., New York, N. Y., 1978.
2. Megill, Robert E., "An Introduction to Risk Analysis," Petroleum Publishing Co., Tulsa, Okla., 1977.

Acknowledgments

Every author incurs debts of assistance in his publications. In the first edition of "An Introduction to Exploration Economics" the author is indebted to Messrs. Paul Costa, Frank Sisti and G. Rogge Marsh for substantial contributions. In addition, Messrs. Frank Holmesly, Harold Ranzau and Fred Heidner read selected chapters and provided their insights. Mr. R. W. Garrett and Mr. D. P. Meagher read the entire first manuscript and their comments were appreciated. To Mrs. Vera Dunlop and all the others who helped go my sincere thanks.

The revised edition of Exploration Economics owes new expressions of gratitude to Mr. Bob Dreyling, Mr. Trent Meadors and Mr. E. C. Capen, who made significant suggestions.

For both editions my largest debt belongs to Mr. G. Rogge Marsh. Chapters 8, 9 and Appendix D owe much to his ideas and thoughts. His constant encouragement was sometimes the necessary ingredient when the burdens of writing seemed insurmountable.

Again it is my sincere hope that each person who contributed to either edition will feel his efforts were rewarded by this book.

R. E. Megill

1 *Setting the Stage*

If you want to know something about the practical economics of the search for hydrocarbons, the pages which follow will provide a beginning. To become an expert in the analysis of exploratory opportunities will require much more reading—and some practical experience in evaluating plays and prospects. One can, however, acquire new knowledge from the writings and experiences of others, accepting or rejecting parts as desired.

Objectives To transmit new knowledge by means of the written word, one must have reasonable objectives in mind. The objectives sought here are:

1. To explain the expanding role of economics in the life of the explorationist.
2. To develop a knowledgeable vocabulary of the terms associated with economic analysis.
3. To review the procedures used in preparing economic analyses, and
4. To relate the new vocabulary and knowledge to some specific analysis.

In general the objectives, if achieved, will do more to help you understand economic evaluations than to enable you to prepare them. You should learn when such evaluations can help you; and having requested an evaluation, you will better understand what is received.

We will mention the role of the computer in today's evaluation methods. Its role is a vital one. Much of what is done today in investment evaluation would be impossible without the computer. Nevertheless, you need to know nothing about computers to derive benefits from this manual. Furthermore, none of the problems requires a computer for solution.

Finally, many of the concepts discussed are common to any type of investment analysis. You will find applications to any area of your life where financial decisions are required.

Your working environment Whether you work for a major, complex organization, in a business partnership or are self-employed, economics is in your future.

If you make your livelihood searching for hydrocarbons, this statement is especially true. The explorationist's work begins at the raw material end of a long investment chain (which ends with gasoline in his tank). If his efforts are successful, profits are made possible for many other "downstream" investments. The exploratory efforts conducted within a major integrated oil company provide investment opportunities for other departments nearer the product end of the investment chain. To begin with, think of your work as contributing to the investment chain.

The first sale of hydrocarbons comes after the development of a property. Thus, the economic success of an exploration program can only be determined after its discoveries are developed and put on production. Exploration economics is in fact the financial consequences of both exploration and production investments—but is related to the time of the exploratory decision.

In a large oil company, the operating departments (exploration, production, marketing, refining, etc.) make the major capital investments. Competition for funds exists between departments. Each department wants money to make investments within its jurisdiction.

Selecting the best investment opportunity With competition for funds, the selection of the best investments becomes a constant process for management. The selection is not always easy. Investments in different departments have:

1. Different tax considerations.
2. Varying degrees of uncertainty and risk.
3. Interfunctional differences in capital requirements, cash flow characteristics, et cetera.

Management decisions involving investment opportunities are influenced by various analysis groups. In a few large companies, most functions have an analysis group within their department. In other companies analysis work is completely centralized. Regardless of the location, these groups provide background evaluations for budget purposes and as an added tool for investment decision making. In any business, large or small, selecting the right investments is the key to growth and prosperity. The principles reviewed in this book should help you select the best available investments.

The decade ahead for the oil and gas industry A brief look ahead will serve to illustrate why economic considerations are in your future.

The problems of the next decade will impose needs for careful scrutiny of all investments—including exploration investments. Many of the factors influencing future investments are beyond the control of the explorationist. Yet, some are so critical that their economic consequence must continually be tested. Let us consider several factors.

1. Interfuel competition

Because the consumer constantly searches for the best product at the lowest cost, we have interfuel competition. Current projections indicate continued shortages of domestic supplies of crude oil and natural gas. Even in situations of tight supply, the competition between fuels will still exist. The competition between fuels *places a ceiling* on prices. The ceiling is a *relative* one which would be affected by general price changes. Consider some of the alternate fuels for crude oil and natural gas.

a. *Shale oil*

Potential shale oil competition is of particular interest. Abundant supplies are available (meaning that no risk is involved in the search). The technology is already available to obtain petroleum products from the kerogen in oil shales of the continental U. S. Crude oil prices might rise under conditions of extremely tight supply. But even if they do, the price can rise only to the point where shale oil becomes an attractive alternate investment. As of now commercial shale oil production is sensitive to:

- crude price
- mining and retorting costs
- tax considerations

- governmental factors affecting the policy on public lands
- disposal of shale wastes

At some time in the future, the right combination of change in these factors will make shale oil deposits economic. At present, the uncertainties associated with each factor preclude an accurate estimate of when this will occur.

b. *Tar sands*

Much the same logic applies to tar sands. Although uneconomic today, tar sands represent potential sources for petroleum products. The deposits being mined in Canada indicate that the technology is available even if the profits, at present, are not.

Up to 1973 the Canadian tar sand operations produced little if any profits. Balance sheets showed lots of red ink! The increased prices for crude oil following the 1973 oil embargo produced profits from the tar sands for the first time.

Tar sand deposits in the U. S. are not as rich as those in Canada. They will require even higher crude prices to be economic.

c. *Nuclear fuels*

Nuclear fuels have entered the competitive energy market through power generation. They should supply an increasing share of future energy for generation of electricity even though present installations appear marginally economic. Political, ecological and economic problems beset immediate expansion in nuclear fuel use. Calefaction of rivers, disposal of waste products, and public concern over radioactive material have contributed to a reduction in the pace of construction of nuclear-fueled plants. Nevertheless, this fuel will become increasingly important. The tremendous energy requirements of the future will cause all fuel sources to expand.

d. *Coal*

We must never forget that both gasoline and natural gas can be manufactured from coal. Thus, the price of this conversion affects the ceiling on future oil and gas prices.

As in the case of nuclear fuels the future of coal expansion is related to more than economic factors. Air quality requirements, wildcat strikes of coal miners, restrictions relating to strip mining and land withdrawal will affect the quantity of future supplies. The longest strike of coal miners in U. S. history ended in 1978. In shifting the major emphasis in energy resources to coal, significant problems in supply could occur. As indicated by the 1977–1978 coal strike, the problem might be one of continuity of mining, not availability of resource.

e. *End-use competition*

The petroleum industry faces end-use competition which reflects back to basic raw materials. The electric car could affect the oil industry since there are several alternate fuels for generating electricity.

Interfuel competition will limit the price charged for any one fuel source. The "ceiling" will be the price of the most reasonable alternate fuel.

2. The trend toward more hostile environments

One of the dramatic changes in the petroleum industry during the past decade has been the discovery of hydrocarbons in very hostile environments. These environments exist from the Arctic climate of northern Alaska to depths of 1,000 ft or greater in the ocean off California. The effect of these environments is to magnify risk and to increase costs; and these two factors mean that economic analyses are vital to proper managerial choices. In the inland United States, the trend toward greater drilling depths produces similar requirements.

a. *The effect of depth*

Depth alone produces significant changes. For example:

- Higher drilling costs
- Costs increase exponentially with depth even for a normal trouble-free well. With depth, however, come increasing chances for mechanical problems, all of which add to cost.
- Well density is less at depth. This factor reduces available information about potential reservoirs. Thus, risks increase with the uncertainties regarding reservoir quality.

The trend toward greater producing depths is apparent from the data plotted on Fig. 1. Each point reflects a new depth record for production. Although there were significant changes in depth of production in 1938 and 1956, the trend toward deeper production has been almost linear for the past 40 years. One need only extrapolate this trend to estimate possible future producing depths. To date, the depth limitation has been drilling technology, not lagging interest in deep prospects on the part of the geologist.

b. *Cost increases not associated with depth*

Significant cost increases not associated with depth are also occurring. For example, geophysical costs are increasing. Crew costs have increased because of complex shooting patterns, common-depth-point procedures, increasing material and personnel costs, and increases in the right-of-way (permit) costs. The computer processing of digital seismic data is another factor of higher cost.

Increases in acreage costs have occurred in the past decade. An increasing proportion of lease bonuses involves sealed bid sales. These sales place the bidder at the disadvantage of knowing only his own bid—in contrast to the auction bid or negotiated price under competition. They produce greater revenue for the seller but represent a higher cost to the buyer. We would not be concerned about higher bonus costs if

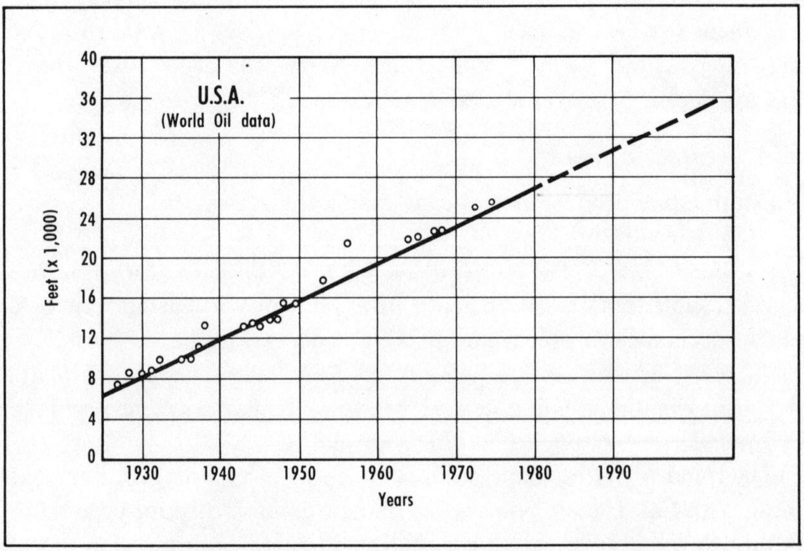

FIG. 1 DEEPEST U. S. PRODUCTIVE PAY

profitable bids were assured. Such is not the case, however. In a sealed bid sale, the arithmetic mean of all bids on a single tract probably represents a profitable estimate of pre-drilling value (which may have little relationship to post-drilling value). Because the highest bid always succeeds, the odds favor an overall marginal profit.[1, 2]

Other acreage costs are also changing. Higher rental payments, shorter lease terms and higher royalty requirements all add to costs.

3. Governmental controls

A third element in cost increases involves regulation, particularly Federal regulation. Government regulation covers such important areas as lease sales, pollution control, pipeline approval (safety and right-of-way) and price regulation (or influence).

The first edition of Exploration Economics recognized the impending changes in industry-government relations. In the ensuing eight years the pace of change has been more rapid than anticipated. In fact most of the changes affecting the petroleum industry since 1970 have been political, regulatory or legislative in nature.

The following paragraphs briefly describe the significant changes since 1970. Both beneficial and adverse changes have taken place. The "good" changes will be reviewed first.

The beneficial changes 1. *Price.* One of the beneficial changes has been increases in the wellhead price for both crude oil and natural gas. Although still not reflecting free market conditions, the controlled price has been allowed to increase for new supplies of both natural gas and crude oil. "Old" gas and "old" oil are still controlled at levels below the market price.

As a consequence of price changes, exploratory and development activities in the late seventies are at the highest levels in a decade. Seismic crews are up, wells drilled are increasing and acreage acquisitions have increased. The industry has responded to the price incentive.

Until 1978 the intrastate gas market had escaped control. It provided an interesting comparison to the controlled market of interstate gas. The supplies of natural gas for the interstate market have declined sharply in recent years as suppliers sought the higher price and unregulated markets in intrastate. In Texas in one recent year, surplus gas existed for a short period as supplies temporarily exceeded the market need.

The increase in wellhead prices has been beneficial to the industry.

2. *Prorationing is gone.* This change is beneficial in that the near capacity production of state allowables has shortened payouts and thus increased cash flow. Well allowables, based on spacing and depth, still exist—so do determinations of MER (maximum efficient rate) for reservoirs. However, since the early 1970's the oil and gas fields of the U. S. have been producing at capacity. There are still "take or pay" conditions in contracts for natural gas, but in most instances these terms apply to few fields today.

3. *Corporate income tax is lowered.* In 1978 the Federal tax rate on corporate income was reduced from 48 to 46%. This slight reduction will not influence exploratory decisions but it does free modestly more cash for reinvestment by the petroleum industry. The tax reduction is one of the good changes.

The good result from these changes is the industry's ability to produce small discoveries previously considered uneconomic.

The adverse changes Many of the significant changes will ultimately stifle the increased activity. They represent the adverse changes since 1970.

1. *The most significant negative change has been in pervasive regulation,* particularly at the Federal level. In 1970 the environmental impact statement (EIS) had not been born nor had many agencies which now regulate oil and gas exploration. Furthermore, the number of agencies contributing to additional paperwork and delay grow more numerous every day. Anyone in the petroleum industry will recognize acronyms of the new entities—DOE, FEA, NEPA, OSHA, NOIAA, FERC, ERA and many more. Some of these agencies are so new that their role relative to the industry is still emerging. In a few cases, FEA for example, the legislative act assigned an almost impossible task—price regulation. Few congressional legislators knew of the complexities of the petroleum industry. Few still do today. As a result a poorly written law, vague and ambiguous, produced numerous illogical rulings which represent compromises of conflicting views. The prime example of illogical regulatory rulings is the "entitlements" concept promulgated by FEA. In effect successful explorers have to pay others to use *their own crude oil*—which is regulated at an arbitrarily low price. Consequences of years of price control for interstate sales of natural gas produced no learning for legislators. The simplest laws of economics show that price regulation produces scarcity, not plenty.

Of course, the recipients of entitlements, the small refiners and east coast refiners, do not see them as onerous. One man's penalty given to another is a welcome bounty.

The end result of the new regulation for industry is overlapping duplicity and mountains of paperwork. Massaging paper neither finds nor develops new energy sources.

2. *Although new crude oil discoveries have favorable prices,* existing fields (in general, as of 1972) have a controlled price roughly half the market price. It is not a simple price control. Even new fields, although eligible for higher prices, have a price ceiling. So the control is not simple, but from industry's viewpoint *absurd.*

Crude oil is now sold at several different price levels. Two levels, near each other in dollar value and affected by market conditions, are unregulated. They are production from stripper wells and the crude oil from foreign imports. (Crude oil and product imports are subject to a small import fee.) The balance of U. S. crude oil production has two lower price ceilings. The lower tier is the "old oil" price, in 1978 slightly more than $5.40 per barrel. The upper tier is the "new oil" price and in general relates to new oil produced after 1972; it totals about $11.50 per barrel.

The present conditions of price control for crude oil emerged after the 1973 embargo. The embargo is over, but the "patient" is still wrapped in his original bandages!

What sort of results would you expect from such regulated pricing? Uneconomic results automatically ensue. Why? Investments tend toward higher profits. If a farmer gets lower prices in one field because of regulation he will concentrate on the fields with better prices. It's common sense! The investor is a rational being, putting his money where the return is best.

Given two identical leases, one controlled at the lower tier price and the other at the upper tier price, on which would you invest money if further development was possible? If you don't develop the lease affected by the upper tier price first, your stockholders should ask for your resignation.

Price controls produce results related to regulation, not to market conditions or real economic value. Controls create a pseudo-economic, illogical non-reality.

The petroleum industry is not the only industry whose costs are increasing markedly from regulatory edicts. Even the academic world, the small businessman, and the farmer are finding their overhead caused by paperwork rising at a rapid rate. Furthermore, despite occasional recognition of the problem by legislative bodies, the number and complexity of regulations increase daily. Delay results and energy supplies are lost or deferred.

3. *Changes in tax laws have also occurred.* Percentage depletion is gone for new discoveries for all but the smallest companies. Furthermore, even the depletion for small operators is being phased downward—to 15% in 1984. The effect of the loss reduces the industry's cash flow at the very time it is borrowing increasing sums of money to cover higher costs of exploration and development of new supplies.

At market conditions the price would have adjusted to reflect the loss of depletion. However, in a controlled market, particularly "old oil," price cannot increase to compensate for the loss. As a result of both depletion loss and the lower price ceiling a few regulated fields in California face premature abandonment. Over-regulation breeds confusion and uncertainty and increased risks.

4. *Inflation has increased substantially.* The rate of inflation for the petroleum industry in 1974 and 1975 was the highest in recent history and much higher than the increases reflected in the common indexes used by government to reflect overall inflation.

Profits, to invest in the search for new supplies, are being eroded by inflation, cost increases from deeper wells and the shift to the hostile environments of the outer continental shelf (OCS).

5. *Finally there exists at this writing an unsettled issue* regarding the basic accounting method to apply to the industry. Most companies, big and small, use what is called "successful efforts" accounting in which losses are deducted from profits in the year in which the losses occur. Some small companies, particularly those just starting, prefer "full-cost accounting." In this method exploration costs, mostly losses, are amortized over a number of years rather than deducted on a current basis. Until this matter is settled in the courts, small operators face uncertainty in obtaining financing.

In addition the Securities Exchange Commission (SEC) has ruled that each oil and gas company must publish an estimate of the value of its reserves. This sharp departure from past practice represents good theory (or intent) but limited understanding of the complexities involved. Numerous projections and estimates of future prices, costs, inflation rates, etc., will produce many interpretive problems. Value is partly a function of one's view of the future and can reflect optimism or pessimism.

Uncertainty is the one word which described the changes which have occurred since 1970. Government policy represents a new and fluctuating and unpredictable variable. One rule, regulation or stroke of a pen can wipe out a forecast or "unhinge" all evaluations.

So, there have been good changes and bad changes. Activity is up, so at the moment the good outweighs the bad but the uncertainty produced by over-regulation, inflation and changing laws looms ever larger as a limiting factor to the growth in new energy supplies.

Whatever their view about the changes, most thought leaders in the U. S. now fully comprehend the vital role that energy plays in the national well-being—indeed to life itself!

The skillful selection of investments necessitates an understanding of all factors inhibiting profits. Government regulation is one of the most difficult factors to predict. Yet its influence increases every year. The petroleum analyst must weigh all factors affecting a proposed recommendation. Even though many of these factors are beyond his control, the explorationist nevertheless must consider them and be cognizant of their significance to his proposals. Exploration economics must include more than just exploratory costs.

REVIEW

1. The future of any business depends upon the proper selections from possible investments.
2. Increasing costs in all segments of exploration, deeper drilling depths, new hostile operating environments and more intensive state and Federal regulation mean that economics is in your future.

BIBLIOGRAPHY

1. Capen, E. C., Clapp, R. V., and Campbell, W. M., "Competitive Bidding in High Risk Situations," paper given at 45th Annual Fall Meeting of SPE of AIME, October 1970.
2. Megill, R. E., "An Introduction To Risk Analysis," Petroleum Publishing Co., Tulsa, Oklahoma, 1977, pg. 164.

2 *Regulation–*
A Dynamic Force

The petroleum industry was essentially an unregulated industry until the 1930's. The discovery of East Texas field in 1930 caused an oversupply problem of unprecedented magnitude. To prevent utter chaos in a vital industry and to prevent physical waste of a natural resource, the various state governments established regulatory agencies.

At first their principal task was to prevent the waste of hydrocarbons and to insure each producer a pro rata share of production. As the industry evolved into a more mature segment of business, other considerations ensued. For a thorough review of modern state conservation read "A Study of Conservation of Oil & Gas in the United States," published in 1964 by the Interstate Oil Compact Commission.

In 1970 the U. S. was trending toward capacity production in oil and natural gas. The nation's production of both oil and natural gas peaked in the early 1970's. As of 1978 the petroleum industry in the U. S. is producing crude oil essentially at capacity and is near capacity for production of natural gas. As capacity production continues imports or alternate energy sources (shale oil, coal, nuclear fuels) must fill the energy gap.

One might think with no surplus capacity existing that regulation would have diminished. Quite the opposite has occurred. Regulation of the oil and gas industry has intensified and become even more of an economic factor. Regulation of various kinds will always be with us; thus, a general review of its significance follows.

The chaos in a vital industry and the loss and waste of the states' resources (and thereby revenue) focused attention on the need for control.

Thus crude oil proration was born. Proration is the restriction of production based on the market needs of purchasers of oil and gas. To administer proration it was necessary to have regulatory bodies. To the oil producing states, who first answered the need for control, fell the design of regulatory agencies.

State agencies In most states regulation is primarily the function of a single agency. These agencies stemmed from conservation laws passed by state legislatures.

It fell their task to determine the immediate problems and to devise adequate regulations for the proper administration of conservation statutes. The name of the dominant regulatory body for several states is listed below:

Alaska	Dept. of Natural Resources
Oklahoma & Kansas	Corporation Commission
New Mexico	Oil Conservation Division
California	Department of Oil & Gas
Louisiana	Louisiana Oil Conservation Commission
Texas	Railroad Commission

Regulation in Texas Because Texas is such an important oil producer, a brief discussion of regulatory activity in that state follows.

A review of regulation in Texas will not be absolutely typical of regulation in other states; nevertheless, something of the broad aspects of regulation, conservation and proration common to all producing states can be learned by reviewing a single state. In addition Texas is a producer of sizable quantities of both oil and gas; as such its regulatory bodies have faced most of the problems of conservation.

In Texas the Railroad Commission (RRC) supervises the regulation of the oil and gas industry. A few activities, involving state land, are regulated by the General Land Office. This body determines the timing and extent of state lease sales. It must approve unitization or pooling agreements involving state lands.

Another state agency, the Texas Department of Water Resources, regulates the industry with regard to the protection of fresh-water sands. This agency approves cementing and casing programs for wells to be drilled including those drilled to dispose of salt water. However, in general, the Texas Railroad Commission has the primary responsibility for regulating most aspects of the oil and gas industry.

The Railroad Commission is composed of three commissioners, elected at large. Current commissioners recognize economic as well as physical waste. It was not always so. Texas' oil and gas conservation policies were created out of the necessity to stop wasteful practices rampant in the industry. Current policies have evolved with the increasing knowledge about reservoir behavior, political pressure, changing economic conditions, supply and demand considerations, court rulings, changing technology, changing environmental concepts, and a maturing industry. Present policies are a curious mixture of logic, tradition and rules of thumb.

What does the Railroad Commission regulate? Its responsibilities cover many aspects of the petroleum industry. The chief responsibilities of the commission are as follows:

1. To regulate the quantity of oil or gas flowing from a tract.
2. To determine well spacing.
3. To authorize or force pooling of acreage (interests) for conservation purposes.
4. To exercise pollution control over production and transportation activities by establishing adequate safeguards.
5. To set conservation standards for and to regulate additional recovery projects.
6. To formulate proration policies.

Some common terms A small, but special, vocabulary relates to the regulation of oil and gas fields. Some terms are common to both oil and gas fields, while others are not. The following terms apply to both oil and gas fields.

1. *Field (common terminology)*—The general area underlain by one or more reservoirs formed by a common structural or stratigraphic feature is commonly called an oil or gas field. *Example:* Alazan, North field.
2. *Field (RRC designation)*—The RRC uses the above field definition when referring to *all* the reservoirs in a field area, but *also* uses the term to refer to a *single* reservoir (commonly called RRC field): *Example:* Alazan, North (H–21 Sand) field.
3. *Well spacing*—Spacing rules regulate the number and location of wells completed in the *same reservoir.*

   The regulations are normally stated as minimum permissible distances from lease lines and between wells. They essentially force each well to drain a specific number of productive acres (proration unit) within the operator's own lease. *Example:* 467′–1200′ well spacing (which means 467′ minimum from the lease line and 1200′ between wells) results in 40-acre proration units.

4. *Proration unit*—The proration unit regulates the number of wells completed in the same reservoir (RRC field).

   The regulations usually state the number of acres which must be assigned to each well. This rule establishes a pattern of well density for adequate drainage of the reservoir and thereby eliminates the drilling of unnecessary wells.

5. *Allowable allocation formulae*—are methods for dividing the total allowable production of an oil or gas reservoir among the operators of the various leases in a manner *considered equitable* by the RRC. *Examples:* Allocation formulas are based 100% on productive acreage (*most common*), on number of wells, ⅓ well—⅔ acreage, acre-ft and so forth.
6. *Field rules*—are those set forth by the RRC to insure orderly development and exploitation in a specific field. They may apply to *one, several* or all RRC fields (reservoirs) in a general area. They usually include spacing rules, density rules (proration units) and an allowable allocation formula—but may include special or unique provisions such as a gas-oil ratio limit or casing requirements. Field rules are normally requested by one or more operators in an open hearing before the Railroad Commission.
7. *Statewide rules*—are a set of general rules applied to all fields (reservoirs) unless specifically amended by field rules. Certain of the statewide rules apply to all fields, including those with field rules. An example is Rule 37 which is the statewide spacing rule.
8. *Exception to statewide Rule 37*—This rule allows flexibility in locating wells in conflict with existing spacing rules.

Terms relating to oil proration Additional terms relate specifically to oil wells and the production prorated to each well. The symbols or letters in parentheses are those used in RRC publications on allowables. The terms are:

1. Top zone allowable (TZA)—is the maximum *per well* allowable granted to completions in a specific reservoir. It is most often based on *depth yardsticks,* but with *many* exceptions.

2. Depth yardsticks—are allowables established by the RRC for statewide application to oil completions. They increase with *depth* and *spacing*. See Appendix A, Table 1, for a summary of allowables by depth yardsticks.
   A. 1947 yardsticks—apply to RRC fields (reservoirs) discovered prior to 1965.
   B. 1965 yardsticks—include greater incentives for deeper drilling and wider spacing. They apply to RRC fields discovered since 1965. Since by definition a reservoir is a RRC field 1965 yardsticks apply to new reservoirs in old fields.
3. Penalized allowables—instead of a TZA, some wells have penalized allowables. Two such allowables are:
   A. High gas-oil ratio (*)—Well production is penalized for casinghead gas production in excess of gas-oil ratio limits to conserve reservoir energy. The daily gas limit for each well is determined by multiplying the TZA by the penalty GOR (usually 2000 cubic feet per barrel).
   B. Limited capacity (#)—is an allowable fixed below TZA because of inability to demonstrate productive capacity.
4. A scheduled allowable—is the top zone or penalized allowable of a well as it appears on the RRC schedule.
5. A proration factor (or market demand factor)—is the percentage of its scheduled allowable that a well is permitted to produce on a *calendar-day* basis. Since April 1972, this factor has been 100% for all fields. The Railroad Commission can reduce the factor in a field to prevent waste or for the entire state if demand for crude oil is lower than supply.
6. Exempt allowables—are not subject to market variation. (In the case of an exempt allowable, the scheduled allowable is the *same* as the calendar-day allowable.)
   A. A discovery allowable (NPX)—is a higher allowable than the standard depth yardstick. The NPX allowable is assigned to onshore wells in a new field until the eleventh well is drilled or for 24 months, whichever occurs first.
   B. A marginal allowable (M)—is used to prevent premature abandonment of economically marginal wells.
   C. Others (X)—allowables may be granted for various reasons.

Terms relating to gas proration Gas production is also regulated by state agencies. State regulation of gas wells concerns the production from the well, well spacing, conservation, pollution control and safety.

Oil well allowables are related to a depth and acreage yardstick. Gas well allowables are initially related to the absolute open flow potential. Later they are related to market demand for gas from each field.

In Texas some of the special terms associated with gas well allowables are listed below:

1. *The absolute open flow potential* (*AOFP*)—is a theoretical value. It is the rate at which a well could flow at the sand face at zero pounds of pressure. The AOFP is a function of reservoir pressure, the product of permeability times thickness (kh), the degree of formation damage and other physical factors.

   Statewide rules require that the AOFP be determined at initial completion of the well and adjusted periodically as the reservoir pressure declines.
2. *Non-associated gas well allowables:*
   A. For fields without field rules, statewide rule 29 fixes the allowable at 25% of the AOFP. Exceptions can be obtained in late stages of depletion to prevent waste or loss of production. An example would be a well loading up with water when produced at a low rate.

   B. *Field rules*—Where field rules exist, the total field allowable is determined from the demand for gas from the field. The total field allowable, as determined, is divided among the wells based on the allocation formula.

   Special allowables to provide flexibility have been granted in fields having a single operator and a single royalty interest—providing a demand exists for additional gas. Multiple pay zones may or may not require development as separate fields. (Rulings in this area can significantly affect deep drilling economics.)
3. *Associated gas well allowables:*
   A. Statewide rule 49(B) covers allowables for wells in the gas cap of a reservoir. The gas allowable is set equal to the daily gas limit of an oil well plus the reservoir volume equivalent to the same well's oil allowable.

B. Sometimes associated gas reservoirs get special allowables as the oil column nears depletion.

4. Natural gas liquid production (lease condensate and plant liquids) is a function of gas production, and normally limited only by gas allowable, except in a few gas cycling projects. In these projects, permission is usually requested to cycle at a specific rate or to limit liquid production to some maximum volume.

The average explorationist will not need to become familiar with these terms. In most evaluations you should consult the local reservoir engineering group (or production geologists) before setting up a schedule of production from a well or field. These two groups understand the terms and their use and give assistance in this area. The terms are provided and defined herein as a reference. A limited knowledge of the terms is of value, however. Terms related to regulation vary somewhat from state to state. See "Handbook for the Oil & Gas Industry" (available at Southern Microfilm Reports, Inc., Metairie, Louisiana) for terms used in Louisiana. For Oklahoma, read "Regulations of the Oklahoma Corporation Commission, Conservation Division."

At the start of *every* economic evaluation (whether of a prospect, play, or basin), one of the first steps involves a scheduling of production. One must learn therefore what well allowables are applicable to the investment or prospect being evaluated.

Federal regulation As we noted, states bore the initial burdens of regulation. Their interest was conservation of a wasting asset for the benefit of their citizens and for the protection of individual property rights. Within the past two decades, Federal regulation has become increasingly important. It affects both oil and gas production.

If someone in the year 1970 could have predicted the degree of Federal regulation which exists in 1978, his forecast would have been rejected. No one at that time foresaw the almost unbelievable magnitude of Federal regulation. The National Environmental Policy Act had been passed in 1969. The Clean Air Act followed in 1970. These new acts placed unprecedented power in the hands of regulatory agencies. The full impact of these laws has just begun to reach affected industries. Furthermore, these two acts were but harbingers of others to follow. Examples are:

Federal Water Pollution Control Act	—1972
Coastal Zone Management Act	—1972
Coastal Zone Management Act (Amendments)	—1976
Endangered Species Act	—1973
Federal Land Policy Management Act	—1976

The avalance of legislation has not slowed. Pending bills before Congress would increase the scope of regulation. They involve:

national energy policy
marine sanctuaries
Alaska lands conservation
National Diversity Act (affecting land usage)
land withdrawals to form wilderness areas.

Little by little the exploration of mineral resources and the multiple use concept of land are being choked to the point of inhibiting development. Yet our need and consumption of resources continues to increase.

The Department of Energy In 1977 the Department of Energy (DOE) was formed. Its formation was instituted to consolidate under one department several agencies dealing with energy policy, resource development, land usage, etc. At the same time the Federal Power Commission (which controlled and regulated interstate gas pipelines and interstate gas prices) was renamed the Federal Energy Regulatory Commission (FERC). The commission has an ill-defined relationship to portions of the DOE but mostly to the Economic Regulatory Administration—a unit of DOE. FERC is not, however, responsible to the Secretary of the Department of Energy. It is an independent body in the regulatory arena of Washington.

The effect Many claims have surfaced regarding the benefits of these new agencies, regulatory bodies and departments. No matter what the claims, two results stand self-evident—*delay* and *paperwork.*

Of course, *delay* to a non-profit organization, has little meaning. To business and industry it means *loss*—loss of money and opportunity. Added delays and redundant paperwork mean tying up people and assets in activities which produce no new product, no new service and no revenue.

The evidence of the effect on industry (not just the petroleum industry) of the new and often overlapping regulations is just beginning to appear. It is unfortunate that the small operator will be most affected. Regulations are issued in such volume that separate staffs in most companies do little else but monitor new rules and disseminate their effect and requirements to operating groups. The company too small for such a staff can easily find itself violating the law—not knowing of the law's existence.

What does it all mean Life for the explorationist has become much more complex. Any projection of future development, prices, inflation, production, and facility installation must now take into account increasing numbers of rules. Furthermore, such rules and regulations are ever expanding. So, in evaluating an investment opportunity, the analyst must be much more versed in existing and pending law than ever before.

Does your offshore location require an archeological survey? Do you have baseline studies completed for ambient air requirements to secure a permit for a gas plant? Has your rig been permitted to drill in this or that ocean area? Have you filed a working plan for exploratory development on Federal lands? Such are the questions which can impact your estimate of development rate for a new prospect.

In practice the analyst must locate within his own orgnization those persons knowledgeable about each regulatory area. In a major company, the Natural Gas Department will have knowledge about existing wellhead prices and possible pending legislation affecting future prices. The Production Department (or Drilling Department) will know of existing rig or location limitations for your prospects. Some companies have in each department a small environmental affairs group. This group sifts from the voluminous Federal Register the new laws, rules or regulations affecting their function. In the absence of such a group, check your Law Department or legal representative.

The special case of natural gas Sales of natural gas in the interstate market have endured the longest regulatory history for the petroleum industry. The Federal Energy Regulatory Commission (formerly the Federal Power Commission) has specific regulatory responsibilities involving natural gas moving in interstate commerce. Interstate transportation and sale of natural gas are regulated "for protection of consumers." Such regulation involves fixing wellhead prices for natural gas. Prices are fixed by broad areas—

inland nationwide
outer continental shelf
possibly the North Slope of Alaska.

Historically prices for *interstate* gas have been restrictive. The result has been rapid growth in supplies for the previously unregulated *intrastate* market and diminished supplies for the interstate market. New discoveries sought the intrastate market where possible. Contracts there were still negotiated at "arms length" and little regulation or paperwork was involved. The inertia of regulation plagues the FERC, reducing its ability to keep pace with rapidly changing conditions and it will be increasingly difficult for it to do so. In 1978 legislation passed by Congress extended price control by FERC to intrastate sales of natural gas. The history regarding interstate gas supplies does not show this change to be any solution to finding new supplies of natural gas. In the long run the answer may well depend on importation of liquefied natural gas (LNG) or synthetic natural gas (SNG) derived from coal or shale. Both sources would produce gas at much higher prices.

Natural gas from Alaska and possibly Mexico will supplement imports from Canada. Imports of natural gas fall under FERC control. Thus far Mexico and FERC have been unable to agree on an acceptable price for natural gas imported at the Texas-Mexico border.

The Department of the Interior exists as a separate body from other regulatory agencies although it now must coordinate its efforts with the DOE. In some instances (offshore lease bidding methods for example) its proposals are subject to DOE review. Nevertheless, the Department of the Interior (DOI) determines the timing and extent of Federal lease sales. Federal lands, however, involve onshore areas as well as offshore. Sealed bid sales are held for offshore Federal tracts and for government-owned shale oil tracts. However, much Federal land, particularly in the Rocky Mountain states, can be leased for a very nominal filing fee. Texas and Louisiana used to regulate production per well in the Gulf of Mexico. However, the Department of the Interior has assumed regulatory authority over production from offshore leases. It and other departments will assume more and more regulatory responsibility in the decades ahead.

Gas contracts As long as gas supply was abundant, gas contracts were reasonably uniform. Guaranteed takes were spaced over a number of years to insure a steady, adequate supply for the pipeline purchaser. The "normal" contract had such general terms as:

1. A negotiated magnitude of the total reserve contracted.
2. A daily contract quantity (DCQ) usually 1 $\overline{\text{M}}$cfd for each 8 billion cu ft of reserve. The DCQ is usually referred to as the "take."

3. A contracted price. If interstate, it was set by FERC. If intrastate, it was a negotiated price. The contract price is often related to gas quality (Btu content).
4. In intrastate contracts some form of price escalation.
5. A committed area (sometimes with a depth limitation) assigned to the contract which may include new reservoirs or extensions to existing reservoirs.
6. Provisions for reserve redetermination. Such provisions allow for an adjustment of the take.

Present conditions of undersupply have spawned many new contract conditions (there are still scattered locations of oversupply in the former intrastate market). Some of the new conditions include faster takes, higher prices, investment prepayment, early price renegotiation and faster escalation. All of these conditions were much easier to institute in the intrastate market when it was not under FERC control. New contract terms can be expected to evolve rapidly until gas price equals the new Federal ceilings or exceeds the most suitable alternate energy source. Gas contract provisions are vital inputs in any evaluation. The analyst must understand the contract provisions to employ them in his evaluation. Conditions can change rapidly, but there still exist in some areas today "take or pay" provisions in contracts which are being exercised because the gas company cannot, occasionally, take all the gas for which it contracted.

Other regulatory influences Other forces tend to act as "regulatory" influences. Some of these forces are:

1. The President of the United States and the economic advisors of the Executive Branch of the Federal government.
2. State and Federal legislatures.
3. City and County governments. Any drilling within a city will necessitate a review of the city's restrictions. County governments have restrictions relating to heavy loads on local roads.
4. Joint interest owners.
5. Royalty owners.
6. Competitors. Occasionally a competitor will initiate an action out of pattern to normal practice. It may set the pattern for future action. It may also influence future legislation if the practice is of questionable legality.
7. Public opinion (which can be a strong "regulator").

8. Supply-demand expectations. What a company expects to happen to total U. S. (world) supply-demand conditions can affect its operations.

With dwindling oil and gas supplies proration has gone—regulation has not; it has sharply increased. The nation's regulatory bodies find themselves regulating more and more over less and less.

In Figure 2 the ratio of reserves to production for the U. S. is plotted from 1962 through 1977. Gas reserves at Prudhoe Bay, Alaska, are not being produced and have flattened the decline in the ratio for gas. When this major field goes on production the sharp decline will resume. Crude oil production has begun from Prudhoe Bay field but its effect on the ratio will be more evident after 1977.

Other than the North Slope of Alaska, almost all hydrocarbon reservoirs are producing at capacity. Normally at capacity the ratio of reserves to production nears the range of 6 to 9. We should expect these curves to move toward that level with gas lagging behind oil until production begins at Prudhoe Bay. With capacity production regulatory emphasis has shifted toward ecological and political control.

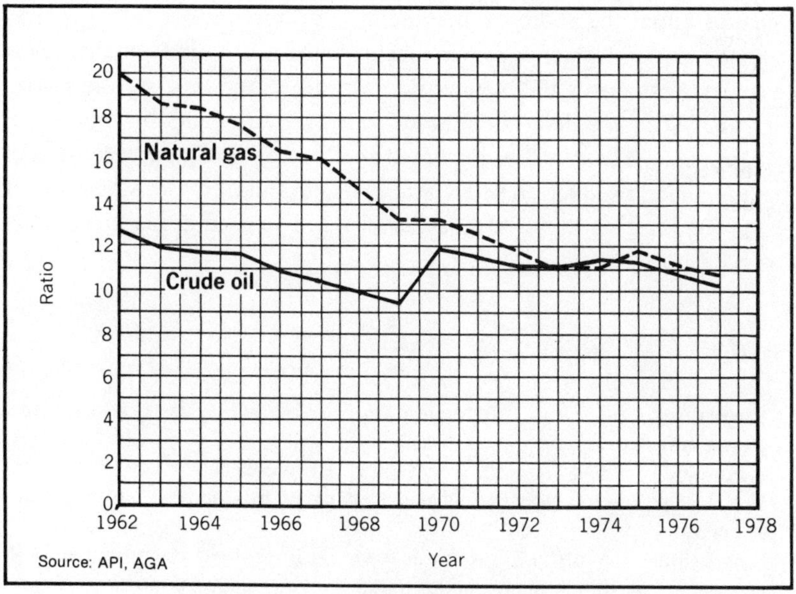

FIG. 2 RATIO OF RESERVES TO PRODUCTION
U. S. SOUTHERN 48 STATES

The shift in regulation The most significant change in regulation has been the subtle shift from regulation in terms of what is lawful, to regulation in how to perform normal business functions. Regulation has moved from an advisory to a participatory role.

Government groups used to say don't do this—or, that is against current rules, as we interpret them. Now more and more groups or agencies tell business, universities, and labor how to do things. This advice is given with authority to punish those who ignore it. The tragedy of the "new" regulation lies in the lack of experience of the regulators. In fact, experience in a complex field is itself suspect and automatically assumed to be non-objective.

What is the result? The result is rapidly growing numbers of employees in Federal agencies and growing inefficiency. Why the growth in numbers and inefficiency? First, it takes more people to tell you how to do it than to tell you not to do it. Secondly, with the experience of regulatory personnel at beginning levels, much learning must take place. Unfortunately the learning experience usually comes at the expense of the segment being regulated.

It is no wonder—no wonder at all—that thoughtful leaders show deep concerns about the ability to operate in the future.

To the explorationist desiring to measure the profitability of a prospect or a play, regulation will be important. It affects future costs and prices. The analyst must determine how, in what way, and by whom his hydrocarbon deposit will be regulated and fit the consequences (cost, well spacing, price, production rate, etc.) into his evaluation. Consideration of regulation is a first step. For this reason it is discussed in the first part of this book.

REVIEW

1. Regulation of oil and gas production was born out of the chaos and waste in a young industry.
2. As the industry has matured, so have regulatory concepts.
3. The U. S. petroleum industry has reached capacity production of its oil and gas reserves.
4. Regulation in the future will continue to emphasize conservation, but there will be even greater emphasis on pollution and environmental control.
5. Because a schedule of production is one of the early steps in the evaluation of an exploration opportunity, a general understanding of regulation is necessary.

3 *Understanding the Terms*

We have progressed to a point where some useful terms must be defined. Exploration expenditures are budgeted, tabulated, and recorded in various records. In this chapter the accounting terms relating to exploration analysis will be introduced. We shall concern ourselves with the type of records normally kept by a corporation. Those with a more simple recording system can adjust accordingly. The terms introduced are the important aspect, rather than a specific recording method.

We will introduce one term at a time by discussion. The terms and associated expenditures will be followed from first investment to final profits. An exploration budget will be developed and budget categories related to the corporate books—which brings us to our first definition.

> *Corporate books:* The method of accounting used to report financial results to management and stockholders.

The method of accounting illustrated in this chapter is the "successful efforts" method in which (most) unsuccessful efforts or losses are deducted in the year spent.

The birth of an idea—we isolate a "lead" Explorationists are familiar with Wallace Pratt's phrase, "oil is found in the minds of men."

Ideas generated by geologists and geophysicists start the investment chain which ultimately leads to the discovery of hydrocarbons. In terms of an exploration budget, idea generation expenditures are represented by the salaries of technical people and have the budget classification of geological and geophysical costs. Salaries and other office expenditures (rent, telephone, supplies, etc.) are usually expensed on accounting records.

Definitions: Expenditures deducted from income in the year of expenditure are said to be *expensed.*

Expenditures deducted from income over the years of useful life of an asset are said to be *capitalized.* Equipment expenditures are usually capitalized. Capital items often have potential salvage value.

In this chapter for each new term the relationship between its budget category and how the item is handled on a corporation's books will be noted. Table 1 shows the budget category, what it is composed of and whether it is expensed or capitalized. This format will be followed for subsequent budget categories.

TABLE 1
Exploration expenditures

	How handled	
Budget category	*Capitalized*	*Expensed*
Geological		Salaries, rent, telephone, etc.
Geophysical		Salaries, supplies, contract services
Management (Overhead and other)		Salaries, supplies, contract services, allocated costs
General plant	Geological warehouse, laboratory equipment; geophysical equipment; office equipment	

"Following up" the idea—a prospect emerges Once an idea emerges, its follow up begins. Additional subsurface maps are drawn and seismic crews set to work. Acreage purchases may ensue. These purchases will involve subsequent rental payments and taxes on unoperated acreage. The terms involved in these activities are defined and the proper budget categories noted in Table 2.

TABLE 2

Exploration expenditures

	How handled	
Budget category	*Capitalized*	*Expensed*
Geophysical crews		Salaries, supplies, contract services
Transportation equipment	Seismic crews, trucks and autos	
General plant	Seismic crews, digital equipment	
Land		Salaries, supplies, rent, etc.
Unoperated lease purchased	Lease bonus	
Rentals and taxes		Rentals and taxes on unoperated acreage
Geological core tests		Core drilling costs

Definitions: A cash payment made to acquire a lease on a property is called a *lease bonus.* A company (or individual) buying the lease is called lessee. The person granting the lease is called the lessor. Since the bonus represents the buyer's investment in the property, it must be capitalized.

The acquisition costs associated with a lease bonus are also capitalized. These costs include:

abstracts of title
recording fees and court costs
certified copies of instruments
attorney's fees
title curative-work
payments to lease brokers

An annual sum paid during the lease life in lieu of development of the property is called a *lease rental.* Lease rentals are *expensed.*

Lease bonus is one of the few *capitalized* budget items which has no salvage value (unless sold to another party before cancellation).

Testing the idea—we get a discovery Assuming that geological and geophysical investigations confirm the original idea, a prospect is born. The prospect can now be tested by several means:

1. A wildcat well can be drilled.
2. A farmout or trade can be made retaining a working interest.
3. A test well contribution can be made to promote an exploratory well.

TABLE 3

Drilling cost analysis

Location ____________ Rig contractor ____________
Lease ____________
Well number ____________
Depth—proposed ________ Drilled ________ Date completed ____
Classification: Oil ____ Gas ____ Drilling days ________

Intangible costs

	Total	Per day	Per foot
Site costs	____	____	____
Transportation—rig	____	____	____
Rig operations	____	____	____
Drilling fluid	____	____	____
Formation evaluation	____	____	____
Casing cement	____	____	____
Completion tools	____	____	____
Perforating	____	____	____
Squeeze cementing	____	____	____
Formation stimulation	____	____	____
Well supplies	____	____	____
Other	____	____	____
Total intangible costs	____	____	____

Tangible costs

	Size	Amount	Cost
Casing—Surface	____	____	____
—Intermediate	____	____	____
—Production	____	____	____
—Tubing	____	____	____
Well equipment	____	____	____
Total tangible costs	____	____	____

Total well cost ________

For the present, assume that we decide to drill a wildcat well. In estimating costs for the proposed well, a drilling cost analysis is usually compiled. Such an analysis appears in Table 3. Note that the total cost is separated into tangible and intangible costs.

The reasons for separating drilling costs into tangible and intangible costs will become apparent in subsequent chapters. For now we need to know that the separation must be made and what items go in each category.

> Definitions: Intangible costs ordinarily do not have a salvage value. They usually account for about 70% of the total cost of an exploratory well.
>
> Tangible well costs ordinarily have some salvage value and average about 30% of total well costs.
>
> Deeper drilling dry hole costs are drilling costs below the deepest producing zone.

Wildcat wells are recorded under the classification of exploratory drilling and appear on budgets as follows:

TABLE 4

Exploration expenditures

	How handled	
Budget category	*Capitalized*	*Expensed*
Exploratory drilling	Successful wells	Dry wells Deeper drilling dry hole cost in successful wells

Note that all costs relating to dry holes are expensed and all costs for successful wells are capitalized—on the corporate books.

A discovery leads to development If the exploratory well is dry, the entire net cost is expensed (deducted from income), and the prospect goes back into the file or additional study ensues before a second well is drilled.

If successful, a chain of new investments begins. More wells are needed to develop the field; lease facilities (flow lines, tank batteries, metering equipment, etc.) must be installed; and the operation of a producing property begins.

Successful development wells are capitalized on the corporate books as are development dry holes. In the case of development dry holes the present handling is a recent change brought about by the Statement of Financial Accounting Standards No. 19 by the Financial Accounting Standards Board.[1] Prior to January 1, 1978, development dry holes were expensed. Successful development wells usually have about 60% intangible costs.

Lease facilities are capitalized. Operating expenses (direct, overhead and benefits) are *expensed.*

Definitions: *Direct operating expenses* are costs identified with specific properties or units of production. They include pumper and gang salaries, repair and maintenance of lease facilities, etc.

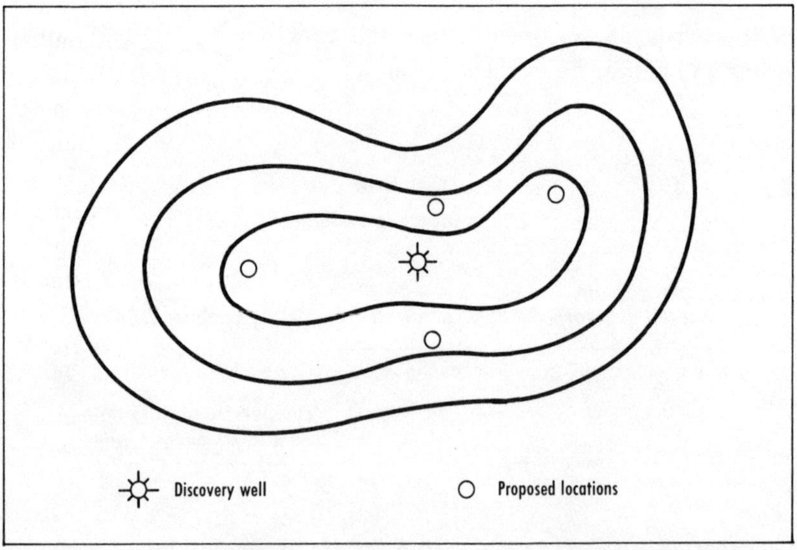

FIG. 3 DEVELOPMENT FOLLOWING DISCOVERY

District overhead is the expense related to operations but which cannot be properly assigned to a specific unit. District overhead is allocated on some rational basis. The term "district" overhead is used to denote the level. It implies operational overhead at the lease level.

In major corporations, there are other overhead allocations for larger organizational units (i.e., divisions, headquarters, etc.).

Employee benefit expense includes all benefit expenditures (vacation, sick leave, holidays, etc.).

Operating expenses begin with production and continue until abandonment. Most expenditures after production starts are expensed. However, additional investments still must be capitalized. Examples of these are:

1. Workover for a multiple completion or recompletion in a new reservoir, if successful.
2. Additional lease facilities; e.g., a compressor for a gas lease when wellhead tubing pressure drops below working line pressure.
3. Additional investment for secondary recovery.

Gas plants are a downstream investment subsequent to discovery and are not exploration investment. These plants are discretionary investments tied to volumes, hydrocarbon composition or special treating needs. Few of these factors are known in a relatively new play; and gas plant economics ordinarily do not greatly affect the yardsticks for judging exploration investments. It is more convenient to assume a plant, if one is needed, and estimate the revenue returning to the leaseholder. After all, you may not build the plant and therefore the plant revenue could go to someone else. For these and other reasons, in this book investment considerations stop at the lease line.

In the event the prospects being evaluated relate to a formation known to carry substantial hydrogen sulfide (H_2S) a plant must be constructed prior to production. In this instance plant investments are calculated and charged to the prospects it serves. The operating expenses for the plant are allocated to the wells served by the plant. These charges usually include a rate of return at or around the minimum acceptable level.

The successful prospect generates revenue The consequences of a successful discovery, its development and its operation are a revenue stream. The revenue is generated by the sale of crude oil, condensate, or natural gas from the lease or income credited to the lease from gas plant products (liquids, sulfur, etc.). A company receives its revenue according to its interest in the property.

Definitions: *Revenue* is defined as the money received from the sale of raw material or products.

A *working interest* (W.I.) is a company's portion (%) of total revenue. It also represents the company's obligation toward investment and operating costs.

A *royalty interest* is a retained interest (usually by the landowner or mineral right holder) deducted from a working interest and having no obligation toward investments or operating costs.

A *net interest* is a company's portion of revenue (W.I.) less all royalty interests.

Tax obligations arise from revenue A profitable producing property creates tax obligations. Out of revenue we pay production, ad valorem, and income taxes. The next chapter is devoted entirely to taxes. They are mentioned here to lead into the calculation of profits from revenue.

How much money does a company keep from the revenue generated by a producing property? What can be deducted from revenue to reduce taxable income legitimately? The items deducted from revenue to reduce taxable income can be divided into two categories—cash and non-cash items. Cash items are those expenditures made during the year and are truly "out-of-pocket" expenses. Non-cash items are deductions allowed from (1) previous expenditures which were capitalized or (2) from depletion. Both types are listed below:

Cash items

1. Direct operating expense
2. District overhead expense
3. Employee benefit costs
4. Production taxes
5. Ad valorem taxes

Non-cash items

1. Depreciation
2. Depletion

Definitions: *Production taxes* are state and municipal taxes levied on oil and gas products, e.g., Texas levies a 4.6% tax on the value of crude oil and condensate and 7.5% on natural gas and natural gas liquids. Production taxes are sometimes called severance taxes.

Ad valorem taxes are state or county taxes assessed upon the value of a property.

Depreciation is an annual reduction of income reflecting the loss in useful value of capitalized investments by reason of wear and tear.

Depletion is a reduction in income reflecting the exhaustion of a mineral deposit. It is allowed to prevent the taxation of a capital asset as ordinary income. The following chapter will illustrate the calculation of depletion.

These various deductions are shown in Table 5. Cash and non-cash items are deducted directly from revenue to calculate taxable income—the amount of money upon which income taxes must be paid.

Table 5 illustrates deductions at the producing leases level. From the total net revenue of all leases, the expenses of higher corporate levels are deducted. Exploration management expenses, research, administrative services, and other allocated costs are examples of such deductions.

TABLE 5

Calculating taxable income at the district level

			M$
Revenue			40,000
Less expense items			
Cash items			
Direct operating expense	3,000		
District overhead expense	1,500		
Employee benefit expense	1,000		
Production and ad valorem taxes	4,000		
Total cash items		9,500	
Non-cash items			
Depreciation	5,000		
Depletion	8,800		
Total non-cash items		13,800	
Total expensed items			23,300
Net taxable income			16,700

A budget system We have now followed the generation of a prospect from its inception through development.

How each expenditure was handled on the corporate books has been indicated, and the calculation of taxable income illustrated. Although corporate usage is the reference point, the general approach to budgeting should be applicable even if you are self-employed. In the final section of this chapter, we will illustrate an exploration budget.

A budgeting system assigns funds for investment purposes. Capital budgeting is needed for an individual, partnership, or corporation:

1. To plan future expenditures.
2. To review the distribution of expenditures by categories, products, geographic areas, geologic trends, etc.
3. To control capital expenditures to make sure that:
   - funds are invested wisely
   - investments meet company standards
   - investments are financially attractive

Some companies budget only capitalized items; others budget all expenditures, including expensed items. If all expenditures in an exploration program are budgeted, the expensed items will dominate. They may account for as much as 70% of the total exploratory budget.

TABLE 6

An exploration budget

____________ Organizational unit
________ Year

Thousands of dollars

Budget category	*Proposed budget*	*Capitalized*	*Expensed*
Exploratory drilling	9,600	2,900	6,700
Lease purchases	4,500	4,500	—
Core tests	300		300
Geophysical crews	3,100		3,100
Rentals and taxes on unoperated acreage	1,200		1,200
Geological expense	1,500		1,500
Geophysical interpretation	1,400		1,400
Land, scouting, EDP *	1,100		1,100
Overhead	2,000		2,000
Equipment **	250	250	
Total budget	24,950	7,650	17,300

* Geological and geophysical data processing.
** Transportation, special geophysical equipment, general plant, etc.

Definition: For accounting purposes the term *investment* includes only those items which are capitalized.

An exploration budget represents expenditures for many separate programs in various stages of development. Some are in the idea stage, some in the "follow up" stage, and others are in the testing stage. In the sample budget shown in Table 6, all expenditures are budgeted. It is symbolic and does not relate to any specific company or organizational unit. It provides a brief review of expensed and capitalized items as shown on the corporate books.

The sample budget tabulated above shows about 30% capitalized items and 70% expensed items. The acquisition costs of acreage leased are included in the budget category of lease purchases. Remember, this representation reflects the corporate books. How we treat these same expenditures for tax purposes is the subject of the next chapter.

REVIEW

1. To illustrate how physical events (wells drilled, acres acquired) are translated into budgets, a prospect was followed from inception to successful completion.
2. Along the way new terms were introduced. We should now understand such terms as expense, capitalize, tangible and intangible.
3. In general for basic accounting records, we capitalize things and expense effort.
4. A company budgets to plan future expenditures, to review the various categories of expenditures and to set up the basis for review, control, and follow-up of investments.

BIBLIOGRAPHY

1. Statement of Financial Accounting Standards No. 19, "Federal Accounting and Reporting by Oil and Gas Producing Companies," Federal Accounting Standards Board, Stamford, Conn., December 1977.

4 *Rendering Unto Caesar*

This chapter is designed to help you understand tax planning. Anyone who searches for hydrocarbons needs an understanding of tax concepts. Whether you are an independent or work for a major company, taxes affect your efforts. If an independent, you probably have a tax consultant and every major oil company has a tax department. Our purpose is not to replace either. It is to increase your awareness of the need for tax planning and to show the effect of taxes on investment evaluation.

The importance of tax planning Taxes are big business. Find out what your company pays in taxes and you will realize just how big the tax business has become. Tax concepts in this chapter are oriented to the corporation and corporate tax rates are assumed. Non-corporate and other taxpayers with different tax rates can adjust the presentations used accordingly.

There are important areas of exploration activity where proper tax planning can reduce the company's tax liability. No one will become a tax expert from reading this chapter. Yet, all of you will be more aware of potential tax savings and how taxes affect an investment.

The current Federal income tax rate for corporations is 46%. For investment evaluation purposes, you may use 50% which allows 4% for state income and franchise taxes. A tax rate of this magnitude means that one dollar of income nets a company only about 50¢. It also means that a dollar of deductions from income saves a company, with profits, about 50¢ of income tax.

With such a heavy burden, a taxpayer must know his rights as well as his duties. Legitimate minimization of taxes (tax avoidance) can be as important as lowering production costs or an increase in sales. In fact a deduction for tax purposes may sometimes be worth several times the same amount in additional sales. For these and many reasons, tax planning makes good sense. Tax planning is the legitimate use of all current laws, regulations, and court decisions concerning taxable income. Tax avoidance is good business planning as opposed to tax evasion which is illegal.

Tax interpretations change periodically from either new legislation or court rulings or Internal Revenue Service practice. To keep informed on significant tax changes and how you are affected, you will need to consult a tax expert; or if you work for a company, check regularly with your tax department.

Important factors in oil and gas taxation Each industry has its own particular types of expenditures which affect taxable income. Since Federal income tax began, various laws, IRS rulings and court decisions have developed regulations governing taxable income. In the petroleum industry, there are six areas which require comprehension for proper tax planning. These are:

1. Intangible development costs (IDC).
2. Percentage depletion.
3. Timing of deductions.
4. Capitalization of expenditures.
5. Capital gains provisions.
6. Investment tax credit.

Of these six factors, the first four are of importance to the explorationist. Capital gains regulations seldom affect exploration investment decisions. The investment tax credit (ITC) is a direct credit (deduction) from income tax based on a percentage of qualified capitalized investments. It has varied from seven to ten percent in recent years and is subject to frequent change by Congress. Thus, in any text it becomes impossible to give finite comments on an indefinite subject. However, for the explorationist, ITC, like capital gains, seldom influences an investment decision.

I. Intangible development costs (IDC)

Intangible development costs are expenditures which ordinarily do not have a salvage value; they must be essential to the drilling and preparation of wells for production of oil and gas. Items making up intangible drilling costs were shown on the drilling cost form (Table 3) in Chapter 3.

For the corporate books, we said that IDC are capitalized and depreciated over the life of producing property. The IRS says that for computing taxable income we may capitalize or expense IDC. Only one choice is possible for each company, and it must be made in the year the company first incurs IDC. Most corporations made the choice many years ago.

Intangible development costs are subject to being reduced if a property is sold. This change was made in the Tax Reform Act of 1976. The tax law states they are subject to "recapture" meaning that, under certain conditions, portions of IDC cannot be claimed and must be reported as ordinary income. Because explorationists normally are not involved in the sale of properties this aspect of tax law is mentioned only to document its existence to protect your tax interest fully. See a tax expert if you sell a property.

A tax savings prevents an outflow of cash in the form of income tax. A company with many investment opportunities may have cash requirements which exceed those generated internally. Capitalizing expenditures defers tax savings and increases current income tax. Expensing expenditures, where allowed, reduces taxable income and thus also reduces Federal income tax. Such a reduction in turn leaves more cash immediately available to be reinvested in new revenue-generating opportunities. In addition a company receives no practical benefits from capitalizing IDC. More comments on this later.

II. Percentage depletion

1. *History*

Percentage depletion came into being as a consequence of the income tax amendment to the U. S. Constitution in 1913. Prior to this amendment, there was no need to distinguish, for tax purposes, between income and capital. Capital expenditures generate revenue from which dividends, profits, and taxes are paid. If capital is taxed as ordinary income, the amount available for re-investment diminishes; therefore, tax laws distinguish between capital and income so that capital is retained for future investments. Consider a U. S. Savings Bond. When you pay $18.75 for a bond, your capital is invested. Later you may cash the bond for $25. The interest (income) of $6.25 is taxed when cashed, but *not* the original investment (capital) of $18.75.

An oil discovery presents some difficult problems in determining its value as an asset. Because of several factors, one of them being risk, a discovery is worth several times the investment in the individual lease on which the discovery well is drilled. To tax all revenue as income would sharply reduce funds for re-investment.

Congress attempted to alleviate the plight of the oil producers with the passage of a law on "discovery value" depletion in 1918. By this law a producer could deduct the fair market value of his property from income to prevent capital loss. Unfortunately, the fair market value of an oil property is not easily determined in the discovery year; and in 1926, the Congress attempted to make the regulations more workable by changing the discovery value concept to a percentage figure. Using the percentage relieved the administrative problems associated with the law and from 1926 to 1969 oil operators were allowed to deduct 27½% of gross income from a property as percentage depletion (with one limitation to be discussed later).

There was nothing magic about the 27½%. It was a compromise between 25% and 30% worked out by a congressional committee. It bore an approximate relationship to the average discovery value which the courts allowed in the cases from 1918 to 1926. In the 1969 Tax Reform Act, the depletion percentage for oil and gas properties was revised downward to 22%.

The Tax Reduction Act of 1975 is a mockery of its title. It raised the taxes of the petroleum industry substantially by virtually eliminating percentage depletion except under special conditions.

For large petroleum companies percentage depletion was eliminated except for:

a. natural gas whose price was regulated by the Federal Power Commission (FPC) and was sold prior to July 1, 1976;
b. natural gas sold before Feb. 1, 1975, under a fixed price which could not be adjusted upward to offset tax increases from the loss of depletion.

For small, non-integrated producers percentage depletion continues at 22% of gross revenue until 1980. As shown by the following schedule it decreases to 15% by 1984.

Year	*Daily average production–bbl**	*Percentage depletion rate–%*
1975	2,000	22
1976	1,800	22
1977	1,600	22
1978	1,400	22
1979	1,200	22
1980	1,000	22
1981	1,000	20
1982	1,000	18
1983	1,000	16
1984 & thereafter	1,000	15

*or natural gas at 6,000 cubic feet per barrel conversion.

The net effect of the Tax Reduction Act of 1975 was to eliminate depletion on oil for large companies and sharply restrict its application on existing gas sales. Percentage depletion for large companies is gone for new discoveries.

Despite the near loss of depletion, it remains a factor, particularly in tax consequences of trades between a large and small company. For this reason the further definition and discussion of percentage depletion and its calculation are retained in this revised edition.

For the layman one of the best references on depletion is a small booklet published by the Mid-Continent Oil & Gas Association. The title is "Percentage Depletion—Economic Progress and National Security." It is recommended reading for those interested in learning more about percentage depletion.

2. *Further definition*

Each producing property has a tax deduction called allowable depletion. Allowable depletion is the amount in dollars which may be deducted from taxable income in a year. It is the greater of cost depletion or percentage depletion.

a. Cost depletion is based on the capitalized cost of a mineral property. For example, if you buy producing oil reserves at $1 per barrel, you can by law deduct $1 from income each time a barrel is produced; the $1 is your capital investment and the law always allows a return of capital, i.e., you are always entitled to cost depletion.

b. Percentage depletion

The discoverer of an oil or gas field did not buy a *producing* property. He recovers his capital through percentage depletion which is 22% of gross income—not to exceed 50% of the net income from the property. The 50% limitation is the one mentioned earlier on page 39. In effect it limits the oil operator to no more than the capital gains provision allows non-mineral investments.

It is not beyond the realm of reason to foresee the reinstatement of percentage depletion at some future date. Why? The tax history of the petroleum industry around the world has been one of "up and down" rates. However, when the needs have been great, incentives have been granted by most governments. Thus the condition of large capital needs and very scarce resources could generate new depletion incentives. Such incentives are needed now for the development of synthetic crude and natural gas resources (shale oil, coal gasification, etc.).

Allowable depletion—its calculation To see how allowable depletion is calculated, we must investigate both cost depletion and percentage depletion.

Cost depletion may be expressed as $I \times \frac{P}{R}$ where

I = The investment (actually the undepleted cost in a property, i.e., the original cost less previously deducted depletion).

P = Units (bls. or Mcf) of oil, gas or liquids produced during the year.

R = Units of oil or gas estimated to be left in the reservoir plus current year production.

The definition could also be shown as:

$$\text{Cost depletion} = \text{remaining investment} \times \frac{\text{annual production}}{\text{remaining reserve}}$$

Table 7 shows a simple calculation of cost depletion.

TABLE 7

Cost depletion

(Cost figures in thousands)

	R		P	P/R			I	P/R × I
	Reserves-MB		MB	Unit of	Lease Cost		Undpl'n.	Cost
Year	*added*	*Net*	*Prod.*	*dpl'n.*	*Added*	*Total*	*cost*	*dpl'n.*
1	500	500	10	.02	100	100	100	2
2	300	790	32	.041	—	100	98	4
3	—	758	58	.077	110	210	204	16
4	—	700	50	.071	—	210	188	13

The first year, from left to right, illustrates the calculation of cost depletion in the year of initial investment. In year 2 reserves are added, the new reserve base is calculated and cost depletion determined. In year 3 additional investment is added and the new investment base is determined, after which cost depletion is calculated. Note that cost depletion is always calculated from the undepleted cost (I). The example shown in Table 7 typifies most depletion calculations for oil and gas leases. Cost depletion usually has meaning only early in the life of a property when the net income is very small or negative. As percentage depletion becomes available with income, it is usually larger than cost depletion and is therefore used instead. More on this later.

3. *Percentage depletion*

Calculating cost depletion is only one of three calculations which must be made to determine allowable depletion. A second step is to calculate 22% of gross income. Since percentage depletion can never exceed 50% of net income from a property, the third calculation is to determine the net income from the property. A simple graph, Fig. 4, illustrates the relationship between cost, percentage depletion, and 50% of net income and how one determines, from these, the allowable depletion.

4. *Calculating net income*

In Table 8 the calculations necessary to find allowable depletion are illustrated. Note the determination of net income. It results from reducing gross income (revenue) by:

a. Operating expense of the lease.
b. Intangible drilling costs (IDC). Remember we said earlier that IDC affected the determination of percentage depletion. See also the notes on IDC, p. 49.
c. Dry hole costs on the lease.
d. Depreciation of lease and well equipment (calculated as a percentage of tangible investment).
e. Overhead for depletion limitation (calculated as a per cent of gross income).

In year 1 the lease net income is negative, thus 50% of income is also negative. Since percentage depletion cannot exceed 50% of net, the 22% of gross cannot be deducted. Assuming we are dealing with the same property as in Table 7, we can only take cost depletion amounting to $2,000. (Remember we can always take cost depletion if to our advantage—see Fig. 4.) Cost depletion is the allowable depletion for year 1 as shown in column 11.

The same condition exists for year 2 where net income is still negative. In the third year, however, with no IDC, the income is positive and we check to see which is the smaller—22% of gross or 50% of net. Since 22% of gross is smaller (but larger than cost depletion), it becomes the allowable depletion.

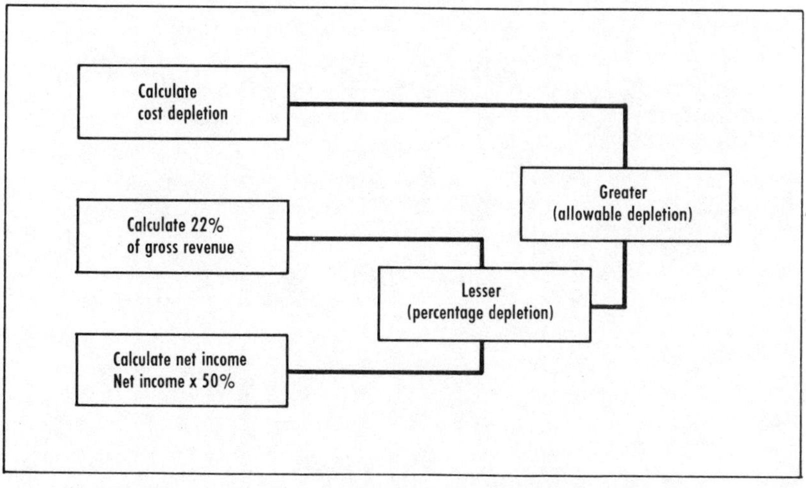

FIG. 4 CALCULATING ALLOWABLE DEPLETION

TABLE 8

Calculating allowable depletion

(Thousands of Dollars)

Yr.	1 Gr. inc.	2 Lse. & Well exp.	3 IDC & Dry holes	4 L&W depr'n.	5 Ovhd. 8%	6 Total exp.	7 Lse. Net inc.	8 50% of Net inc.	9 22% of Gr. inc.	10 Cost dpl'n.	11 Allow. dpl'n.
1	30	7	110	1	2	120	(90)	(45)	7	2	2
2	96	18	100	2	8	128	(32)	(16)	21	4	4
3	175	35	—	4	14	53	122	61	39	16	39
4	150	31	—	3	12	46	104	52	33	13	33

Year 4 is similar to year 3. In both years three and four, 50% of net is greater than percentage depletion, but by law we must take the lesser of these two numbers. Normally, it is only in the last years of a property's life (or under conditions of very high operating costs) that the 22% is restricted by the 50% limitation. Note that columns 8, 9, and 10 contain the three separate calculations which must be compared to determine allowable depletion. Again, see Fig. 4.

III. The timing of deductions

We have already briefly mentioned the importance of timing in tax planning. The timing of a tax deduction refers to the year in which an expenditure may be deducted, if deducted at all. Because of ever present cash needs, a corporation usually tries to get tax deductions at the earliest practical date. Deductions save taxes; and taxes saved are the same as a return of cash to a company or at least an earlier availability of money.

Most companies with taxable income would expense all expenditures for tax purposes if permitted by the IRS. However, some expenditures must be capitalized; but even capitalized expenditures have different rates of depreciation permissible. Therefore, our next discussion will center on the timing of deductions through depreciation.

> Definition: Depreciation is a deduction from income reflecting loss of value of tangible equipment due to wear and tear. It reduces taxable income by charging (depreciating) a part of the cost against each year's income.

Computing depreciation Several depreciation methods are used in determining corporate income and taxable income. These methods can be grouped into three categories:

1. Straight line
2. Accelerated
   a. Declining balance
   b. Sum-of-the-Years-Digits (SYD)
3. Unit of production

The straight line method The straight line method for computing depreciation deducts an equal increment each year over the life of the property.

In calculating straight line depreciation, the depreciable amount is divided by the number of years considered to be the useful life of the property. (In actual accounting practice, it is the depreciable amount less the estimated salvage value. However, in our evaluations we ignore salvage value for reasons which will be more obvious when we study the time value of money. Note, therefore, that the depreciation discussions for all methods are simplified by excluding the effect of salvage value. This simplification will not preclude sound exploratory decisions.) Dividing the depreciable amount by years of useful life yields the annual depreciation deducted from income. For example, assume a depreciable cost of $100 and a 10-year life. The annual depreciation is:

$$\frac{100}{10} = \$10 \text{ per year for 10 years}$$

Accelerated methods Accelerated methods for computing depreciation apportion larger amounts of the depreciable investment to earlier years in the life of the equipment. You get a faster tax write off—an earlier availability of money—which we have already stated is desirable. A discussion of two common methods of accelerated depreciation follows:

1. *The declining balance method*

In the declining balance method, the annual depreciation is calculated at 150% or 200% of the straight line rate applied to the undepreciated balance of the investment. For tax purposes both the 150% and 200% declining balance methods are used. The 200% rate is sometimes called the double declining balance. The 150% factor is applied to used equipment or new non-residential buildings. The 200% factor is used for new equipment.

Assume the same $100 investment and the 10-year life used in the straight line method, but now we will accelerate the depreciation using the double declining balance (200% factor). The straight line method deducted 10% per year. The declining balance method (200%) will deduct 20% per year of the remaining investment. Computations for the first few years are:

First year	$100 × .20 = $20
Second year	($100 − $20) × .20 = $80 × .20 = $16
Third year	($80 − $16) × .20 = $64 [23]× .20 = $13, etc.

2. *Sum-of-the-years-digits method* (*SYD*)

The SYD method applies only to new equipment having a useful life of at least three years. The investment is multiplied each year by the proper SYD fraction. In this fraction the numerator is the remaining life of the equipment, as of the beginning of the year, and the denominator is the sum of the digits in the total life of the equipment.

Assuming the same investment (depreciable cost) of $100 and a 10-year life, the SYD depreciation would be as follows:

Sum-of-the-years-digits = 10 + 9 + 8 + 7 + 6 + 5 + 4 + 3 + 2 + 1 = 55

Year	*SYD fraction*	*Per cent*	*Depreciable cost*	*Computation*	*Depreciation*
1	10/55	18.2	100	100 x .182	$18
2	9/55	16.4	100	100 x .164	16
3	8/55	14.5	100	100 x .145	15...etc.

With the preceding explanations, we are now in a position to compare the accelerated depreciation methods with the straight line method. A comparison follows:

TABLE 9

Comparison of depreciation methods

	Straight line		*Declining balance*		*Sum-of-years-digits*	
Year	*Annual deduction*	*Cumulative*	*Annual deduction*	*Cumulative*	*Annual deduction*	*Cumulative*
1	$ 10	$10	$ 20	$20	$ 18	$18
2	10	20	16	36	16	34
3	10	30	13	49	15	49
4	10	40	10	59	13	62
5	10	50	8	67	11	73
—		—		—		—
6	10	60	7	74	9	82
7	10	70	5	79	7	89
8	10	80	4	83	5	94
9	10	90	3	86	4	98
10	10	100	14	100	2	100
Total	$100		$100		$100	

At the end of the fifth year, the half-way mark, the SYD method has produced 50% more cost recovery than the straight line method. In the same time, the declining balance method (200% factor) has recovered 33% more. Thus, either accelerated method returns capital faster than the straight line method.

The unit-of-production method The unit-of-production method is used to depreciate lease and well equipment that has a life largely controlled by the physical depletion of reserves. It is a similar calculation to that of cost depletion.

The amount to be deducted each year may be expressed as:

$$(C-D)\times\frac{P}{R} \text{ or } (\text{Cost}-\text{Depreciation})\times\frac{\text{Production for year}}{\text{Remaining reserve}}$$

where

C = the cost of equipment
D = accumulated depreciation
P = barrels or Mcf of gas produced during the year
R = recoverable barrels or Mcf remaining in the reservoir at year end plus current year's production

An illustration of the calculation follows:

	C	D	C-D	R	P	$\frac{P}{R}$	$C\text{-}D \times \frac{P}{R}$
Year	*Cost of equip.*	*Prior years accum. deprec.*	*Depreciable balance*	*Reserves*	*Prod.*	*Deprec. rate*	*Annual deprec.*
1	100	—	100	40	2	.050	5
2	100	5	95	38	4	.105	10...etc.

Points to remember about depreciation

1. For tax purposes, you should take advantage of *accelerated* methods whenever possible.
2. You benefit substantially by increasing the rate of depreciation in the earlier years.
3. Computed by any method you get tax deductions equal to the original (depreciable) cost of the item exclusive of salvage.

4. The axiom to remember is—take the biggest deduction possible at the earliest date possible.

IV. The tax treatment of exploratory expenditures

There are certain basic considerations in the potential capitalization of any expenditures for tax purposes. These are:

1. Our first inquiry is always—must the expenditure be capitalized or can it be deducted as current expense.
2. If deductible (expensed) no further tax considerations are necessary.
3. If capitalized we must ask:
   a. Must it be permanently capitalized?
   b. May all or part be deducted for tax purposes in later years?
   c. Will any tax benefits be derived?

As a rule of thumb for tax purposes, we should capitalize as little as possible, expense as much as possible. Of course, the application of this general rule must be within the law.

Lease bonus Lease bonus payments are capitalized for both booked and tax purposes. The tax laws state that the bonus on a productive lease is recovered only through the depletion allowance. For large companies, because of the Tax Reduction Act of 1975, this limits the recovery of bonus to cost depletion only. If the lease is non-productive and surrendered, the bonus is deductible (expensed for tax purposes) in the year of surrender.

An allocated portion of lease bonus must be subtracted each year from gross income before figuring depletion, i.e., gross income for calculating taxable income is not the same as gross income (depletable income) for calculating depletion. The amount per year is usually so small that it is ignored in our evaluations. It obviously cannot be ignored for large bonuses, such as offshore sealed bid sales.

Intangible drilling costs IDC must also be recovered through allowable depletion if capitalized. Herein lies the strongest motivation for expensing intangibles. If capitalized they are recovered, now mostly through cost depletion, but over the life of a property. The time delay in recovery lowers near-term cash flows and therefore is much less desirable than expensing intangibles and getting complete recovery in the year of expenditure. Thus all companies expense IDC as allowed by law.

Geophysical crew costs, test-well contributions (TWC) and core drilling These expenditures are expensed on the corporate books but may be capitalized or expensed for tax purposes. A company usually will expense these costs if possible.

The IRS, however, has ruled that—if the information received adds value or contributes to a decision to acquire or retain acreage, the cost must be capitalized. Divergent opinions are possible in this area and discussions are often held with IRS agents to interpret results from such expenditures. Ordinarily historic experience is used in investment evaluation to estimate capitalized geophysical crew costs. However if these costs are a significant per cent of total exploration costs special adjustments may need to be made. It is the expensed items which receive the most careful scrutiny. Capitalized charges are not questioned by the IRS. In addition to crew costs, some data processing, interpretation (both geological and geophysical) and portions of overhead are now capitalized by IRS practice.

Trades Trades may involve farmouts, farmins, or contributions. Tax planning for trades concerns deductions for IDC and allowable depletion if available to either party.

The tax consequences of trades can be complex. In addition changes occur continuously as a result of legislation, court rulings, or changes in IRS regulations. For example, a recent IRS ruling, (77-176), makes certain types of trades subject to much more income tax. Because the tax consequence of trades vary considerably by virtue of the type of trade and from changing rulings and legislation, they will be discussed only in very general manner here.

Trades vary considerably. In fact the possible variations are almost infinite. However, trades can be divided into three major classifications:

Full assignment of W.I.	*Partial assignment*	*Contribution*
A. With an override retained	A. With a working interest retained	A. Dry hole money
B. Subject to a profit sharing	B. With a convertible override retained	B. Bottomhole money
	C. A carried interest agreement	C. Acreage

The full assignment relieves the assignor of most responsibilities and usually poses the fewest tax problems. Partial assignments, on the other hand, generally pose the most numerous tax problems to the assignee.

The tax consequences of trades require special study. Unless trades are quite uniform, the best advice is to consult your tax advisor or your tax department. An important reference in this area is Arthur Anderson and Company's "Oil and Gas Federal Income Tax Manual."

Tax effects must be considered carefully in trades because profitability can be materially increased by wise tax planning.

Summary As a final resumé of all we have discussed, a table follows in which the tax treatment of exploration expenditures is compared to the

TABLE 10

Summary—tax treatment vs. book treatment

	Tax treatment	*Book treatment*
1. Dry holes—Exploratory	Expense	Expense
—Development	Expense	Capitalize
2. Successful wells		
Intangible well costs	Expense	Capitalize
Tangible well costs	Capitalize	Capitalize
3. Lease bonus	Capitalize	Capitalize
4. G & G costs:		
Core drilling	Capitalize or expense	Expense
Seismic crews	Capitalize or expense	Expense
Data processing	Capitalize or expense	Expense
Interpretation	Capitalize or expense	Expense
Test well conributions	Capitalize or expense	Expense
Portion of overhead	Capitalize or expense	Expense
5. Depreciation:		
Lease and well equipment—new	Sum-of-years-digits	Unit of production
Other equipment—new	Sum-of-years-digits	Straight line
6. Depletion	Percentage or cost depletion	Cost depletion
7. Unoperated leases	Expense (as surrendered)	Amortized
8. Rentals and taxes	Expense	Expense

booked treatment. We think you will find this table to be one of the most useful references in the manual.

The word amortization, item 7 in Table 10, has not been defined herein. It is somewhat akin to depreciation except it does not relate to tangible things which wear out. It relates to intangible items such as surrendered leases. It is a means of deduction, over a period of years, of the value of an expenditure which is not expensed. The amount amortized (deducted) each year is usually based on experience.

REVIEW

The purpose of this chapter has been to isolate important areas of exploratory activity where proper tax treatment can reduce taxes. The subject of Federal income tax has been treated very generally; and no one will become a tax expert from our discussions. We hope, however, that an awareness of potential tax savings has been gained. The explorationist should consider the effect of taxes in analyzing and comparing investment opportunities.

5 Converting Exploration Programs to Cash Flows "Cash Flow Analysis"

Previous chapters dealt with the terms and concepts needed to understand investment evaluation. This understanding of methods and terms was necessary in order to proceed with more complex methods.

Most corporations have growth as an objective. Spectacular growth can be achieved for the small company, either through merger or fortuitous investments. As a corporation gets larger, it takes increasingly larger new investments to insure growth. Thus every major corporation must wisely reinvest part of its profits to enable continued growth. The rate of growth becomes a test of the skill in evaluating and selecting the best new investments from the stockpile of opportunities.

Some companies are capital limited. In such cases the best investments are selected, starting at the top. Investing continues until no funds are left. For a company not capital limited, a different approach applies. In this instance minimum standards are set and all investments are made which qualify. The minimum standard must be above the cost of capital, to insure growth; and a company's rate of growth will be determined by its ability to make a satisfactory return on new investment opportunities. Regardless of the financial ability of a corporation, it is important to compare investments on a common basis.

Large investments in high-risk ventures require careful and sound evaluation. Major corporations have the financial ability to make large investments. They also have, within reason, the capital to invest in high-risk ventures if the potential gains are worth the risk.

One of the basic tools to evaluate and compare investment opportunities is cash flow analysis. This chapter builds the basis for converting exploratory plays or programs into cash flow streams.

Cash flow analysis The end result of matching the inflow and outflow of all related funds, over the life of an investment, is a cash flow stream.

The calculation of a cash flow stream is frequently referred to as building an economic model. Regardless of the terminology, a technique is needed to depict expenditures for investments and subsequent revenue generated. Cash flow analysis not only relates investments to subsequent revenues, but also enables a comprehendable comparison between investments.

The difference between inflows and outflows over a period of time adds up to profit or loss. This difference is called net cash flow. Note that we have said that the net cash flow can be positive or negative. For most investments, particularly those in the exploration department, the early annual cash flows are negative. They are negative because outflow (the money to make the investment) exceeds inflow (revenue).

One can use cash flow analysis for any entity which has both income and expense. A corporation, individual, government, club, or organization can use this principle.

The total cash flow, positive or negative, for a corporation is the sum of the cash flows for all functions. In exploration economics we look at the cash flows for the combined exploration-production investment stream. We try to analyze all opportunities on a standard basis, building our economic models as carefully and realistically as possible. As we learned in the preceding chapter, proper income tax calculations are an important facet of cash flow streams.

Our construction of a mathematical model for exploration-production investments will separate the cash flow into two segments. In so doing the tax effects on each segment can be more clearly shown. We will investigate first the investment cash flow stream (after Federal income tax) and secondly the income cash flow stream (again, after FIT). The sum of the two streams will be the after-tax net cash flow or the profit from the investment.

The investment cash flow stream Before we can set up our investment cash flow stream, we must break our expenditures into sub-groupings for tax purposes. The separation is shown on the following table which categorizes for tax treatment the expenditures associated with the School prospect.

TABLE 11

"Investment" cash flow stream—$M
School prospect

	Year	Capitalized [a]	Tangible [b]	Intangible [c]	Expensed [d]
		Expenditure type for tax treatment			
Expl.	0	500	—	—	—
	1	100	—	—	200
	2	—	—	—	1200
	3	—	200	700	100
					Exploration expenditure = $3,000
Prod.	4	—	500	1500	300
	5	—	100	—	800
					Production expenditure = $3,200
	Total	600	800	2200	2600
		Booked investment: $3,600			Total = $6,200

[a] Lease acquisition and leasing cost, part of geophysical, geological, data processing and division overhead.
[b] Part of successful well costs, lease and service facilities.
[c] Part of successful well costs.
[d] Scouting, lease rentals and taxes, dry holes, overhead on investment and part of leasing, geological, geophysical and exploration overhead.

First let's deal with the exploration expenditures. Our separation isolates tangible and intangible costs for wells, capitalized exploration costs and those items which are expensed. We can follow the development of the School prospect by looking at expenditures, the category in which they are placed and the year of expenditure.

Note that the first year is given the number zero. Year zero is considered, in cash flow analysis, the point of the first significant expenditure associated with the investment being analyzed. One can start a cash flow analysis from any decision point. Downstream decision points, if evaluated, represent partial evaluations only and should be undertaken with the full knowledge that they do not reflect the entire investment. Partial evaluations are sometimes improperly called "incremental" investments. An exploratory investment made to generate a profit is actually a series of investments each requiring a decision based on the outcome of prior investments.

Production department expenditures amount to \$3.2 million, which with exploratory funds yield total out-of-pocket expenditures of \$6.2 million. The capitalized investment, however, (what we show on the corporate books as an asset) amounts to only \$3.6 million.

The investment cash flow stream From the separations made in Table 11, an investment cash flow can be constructed. Remember, we are separating the investment and revenue flow streams only for convenience and ease of explanation. Our investment flow stream for the School prospect is shown in Table 12.

In year zero of Table 12, an expenditure of \$500,000 is indicated under the capitalized column. This number represents lease bonus and the net outflow for year zero is −\$500,000. In year one more leases are purchased, and geophysical work is conducted on the prospect and some exploratory expenses are incurred both adding up to \$200,000.

TABLE 12

"Investment" cash flow stream (after Federal income tax)—\$M
School prospect

		Outflows			*Inflow*	*Net outflow*
	Year	*Capitalized*	*Tangible*	*Total intangible and expenses*	*Income tax credit*	*Investment cash flow (after FIT)*
Expl.	0	−500	—	—	—	−500
Expl.	1	−100	—	−200	+100	−200
Expl.	2	—	—	−1200	+600	−600
Prod.	3	—	−200	−800	+400	−600
Prod.	4	—	−500	−1800	+900	−1400
Prod.	5	—	−100	−800	+400	−500
	Total	−600	−800	−4800	+2400	−3800

Total finding cost = \$3,000,000 (Exploration expenditures)

Net finding cost (AFIT) = \$1,900,000 (Exploration expenditures after tax credits)

Net finding cost (AFIT) = \$3,800,000 (Exploration and production expenditures after tax credits)

Total outflow = \$6,200,000 (Exploration & production expenditures)

Note the income tax credit column. It shows a +\$100,000. Where does this money come from? Remember our tax rate of 50% and the \$200,000 of expensed IDC items. In cash flow analysis, each investment receives credit for the income tax saved. In spending the \$200,000, we incur a reduction in income tax of \$100,000 and the School prospect receives credit for this amount. It shows up, then, as an inflow under the tax credit column. This procedure only has merit if a company has profits from which such deductions can be made. A company with no income could not take such a credit.

Because of the tax credit the net outflow for year two is only \$200,000. We spent \$300,000 but saved \$100,000 in income tax. The government did not give us \$100,000, but we saved \$100,000 in taxes by virtue of the \$200,000 in expensed costs.

In year two the \$1,200,000 in the intangibles and expensed column indicates two exploratory dry holes, but they generated a tax credit (inflow) so that the net outflow is only −\$600,000. Year three represents the last year of exploration investment. The \$200,000 under the tangible column (an outflow) indicates the discovery well. The \$700,000 of intangible drilling costs (see Table 11) and the \$100,000 expense generate together tax credits of \$400,000 making the net outflow for year three \$600,000.

The sum of the net cash outflow to this point represents after-tax exploratory expenditures. They total \$1.9 million. Compared to total exploratory expenditures of \$3.0 million, we see the significance of the tax credits (inflows) to our investment stream. Finding costs expressed as dollars per bbl. or cents per Mcf usually express total expenditures, not the after-tax investment.

An important point is illustrated by the tax credits. When a company ceases to make new investments (fails to incur depreciation and expense deductions), its taxable income and income tax increases sharply. A liquidating company always looks good on the stockholders report because of a high return on booked assets. But the lack of new investments means it is living on borrowed time. One sees here the motivation into diversified investments where a surplus cash flow exists.

Years four and five show the outflows for production department expenditures. These outflows include the wells drilled to develop the prospect plus the necessary lease facilities (tank batteries, gathering lines, separators, etc.).

Tax credits are still taken for savings from expensed items, mostly intangible drilling costs. The after-tax production investment is also $1.9 million, compared to a $3.2 million total expenditure. The sum of the yearly net investment outflow represents, for cash flow analysis, the total after-tax investment—for the School prospect it totals $3.8 million. The $3.8 million plus the tax credits of $2.4 million equal the total expenditure of $6.2 million. Because we have other profits from which we can deduct expense to gain tax credits, our company's out-of-pocket expenditure is only $3.8 million. Again the magnitude of the tax credits illustrates the significant difference between before and after tax investments.

Before leaving the investment cash flow stream, there are several points to be re-emphasized.

1. The investment cash flow stream is most often combined, at the outset, with the income cash flow stream. It was separated here for clarification purposes.
2. Remember the zero year as the year of the first significant expenditure for the investment opportunity. One can push the zero year to any "downstream" investment to test the contribution to net cash flow of that partial investment. Think of the zero year as a point of decision.
3. Unit finding costs are normally expressed in total expenditure dollars, not after tax.
4. The capitalized expenditures of exploration must be separated from tangible investments (which are also capitalized). Tax calculations necessitate the separation. Tangible costs associated with drilling cost, lease facilities, flow lines, etc. will be depreciated over the life of the property. The capitalized lease bonus of non-productive leases is expensed when the leases are cancelled or allowed to expire.
5. Watch for investments occurring later in the project life (major workovers, compression to boost pressure in semi-depleted gas reservoirs, secondary recovery drilling and equipment, etc.).
6. Try to include all real or estimated expenditures associated with an opportunity. Be as realistic as possible.
7. If you work for a major oil or gas company, a centralized group will probably supply guidelines for costs or factors difficult to isolate. If not, you must estimate them on your own. Examples are:

- environmental or permitting delays
- future oil and gas prices
- overhead charges
- tax rates

The income cash flow stream After the successful discovery well and subsequent successful development wells go on production, both inflows and outflows of cash result. The cash inflows are:

1. Gross revenue from the sale of oil, gas, natural gas liquids, or sulfur is the basic inflow. Sometimes there is a small amount of non-depletable income from gas plants not credited to the lease. Such sums can be ignored for exploration-production evaluations in most instances.
2. Tax credits count as inflows, as explained previously. These credits result from:
   - expensed items
   - depletion
   - depreciation
3. Occasionally an investment in one plant reduces expenses in a related or connected facility. Since this reduction represents a true savings the new investment receives credit for the reduction.
4. Net salvage value is also a type of inflow.

Items three and four are mentioned for completeness only. They are usually ignored in exploration evaluations unless the salvage value is a high per cent of the original investment. Ignoring salvage will not ordinarily significantly affect prospect or play economics. If salvage values will affect future cash flows significantly, then the dollar amounts should be entered in the appropriate year of the cash flow stream.

The outflows resulting from a successful discovery are: (Definitions for these terms are in the preceding chapter.)

1. The royalty. It is paid to the farmer or mineral fee holder. Usually the mineral fee holder receives one-eighth the value of all production, sometimes more. Royalties larger than one-fourth are rare. In the special case of bidding for Federal leases on the outer continental shelf, royalty can be the bid variable; or a sliding scale royalty (varying with producing rate) can be coupled with bonus

as the bid variable. The basic problem carried throughout this book will deal with the more general fixed royalty situation. The other two special instances along with net profits bidding will be discussed in Chapter 10.

2. Production and ad valorem taxes. These outflows (expenses) are state and local taxes.
3. Direct operating expenses.
4. Employee benefits.
5. Overhead above the level of lease operations.
6. Federal income tax.

Having listed the income stream inflows and outflows, an example can be presented. An income cash flow stream for the School prospect follows in Table 13.

By setting up column headings, one can easily see the relationship between the inflows and outflows of the income stream. From year zero through year three, no income was generated; remember the discovery

TABLE 13
"Income" cash flow stream—M$
School prospect

	1	*2*	*3*	*4*	*5*	*6*	*7*
Year	*Gross revenue*	*Royalty*	*Net revenue (1-2)*	*Prod. & Ad Val. tax*	*Dir. oper. exp.*	*Ovhd. above lse.**	*Inc. cash flow before FIT (3-4-5-6)*
0-3							
4	2,890	480	2,410	240	60	20	2,090
5	2,890	480	2,410	240	60	20	2,090
6	3,160	530	2,630	260	60	20	2,290
7	3,160	530	2,630	260	60	20	2,290
8	3,160	530	2,630	260	60	20	2,290
9	1,560	260	1,300	130	30	10	1,130
10	1,560	260	1,300	130	30	10	1,130
11	1,560	260	1,300	130	30	10	1,130
12	1,560	260	1,300	130	30	10	1,130
13	1,560	260	1,300	130	30	10	1,130
Total	23,060	3,850	19,210	1,910	450	150	16,700

*Includes employee benefit expense

well was only completed in year three. Gross revenue begins in year four and lasts for 10 years, through year 13.

The net revenue for each year is shown in column three and results from subtracting the royalty from gross revenue. Taxes, direct operating expenses, and overhead are deducted from net revenue to get the income cash flow before Federal income tax, column seven. We started with $23.0 million which nets to $16.7 million before depreciation, depletion and Federal income tax.

From this calculation of income (BFIT), we can proceed to the determination of Federal income tax. Again the computations fall neatly into a few columns, as shown on Table 14.

TABLE 14

"Income" cash flow stream (after Federal income tax)—M$
School prospect

Year	7 *Income cash flow BFIT*	8 *Surr. leases*	9 *Deprn. & depln.**	10 *Taxable income (7-8-9)*	11 *Federal income tax*	12 *Income cash flow AFIT (7-11)*
0-3						
4	2,090	200	110	1,780	890	1,200
5	2,090	200	110	1,780	890	1,200
6	2,290		110	2,180	1,090	1,200
7	2,290		110	2,180	1,090	1,200
8	2,290		110	2,180	1,090	1,200
9	1,130		70	1,060	530	600
10	1,130		70	1,060	530	600
11	1,130		70	1,060	530	600
12	1,130		70	1,060	530	600
13	1,130		70	1,060	530	600
Total	16,700	400	900	15,400	7,700	9,000

*Calculations simplified for illustration purposes

Note the column numbers continue from those on Table 13. The consecutive column numbers will help if you desire to combine the two tables. Column seven is repeated. From income (BFIT) surrendered leases, depletion and depreciation are deducted to arrive at taxable income—column ten. (Surrendered leases were not included in the investment cash flow stream because they are a non-cash item. Cash flow measures *cash changes* only and surrendered leases affect only income tax on a cash basis.) At a 50% tax rate the Federal income tax represents one-half of net income and is $7.7 million. The School prospect does not pay $7.7 million in taxes because of the tax credits shown on Table 12.

Now to get the income cash flow stream, we merely subtract column eleven (Federal income tax) from net income (BFIT), column seven. The result is an income cash flow stream totaling $9 million.

Net cash flow Our cash flow analysis has proceeded from the outflow of investment dollars through revenue and the taxes they generated. We have taken advantage of an investment and have estimated its future income from production sold. When we combine the two streams, we have the net cash flow after Federal income tax (AFIT). The combination is shown on Table 15.

TABLE 15

Net cash flow (After Federal income tax)—$M
School prospect

Year	*"Investment" cash flow*	*"Income" cash flow*	***Net cash flow (after FIT)***	*Cumulative net cash flow*	
0	−500		**−500**	−500	
1	−200		**−200**	−700	
2	−600		**−600**	−1,300	
3	−600		**−600**	−1,900	
4	−1,400	1,200	**−200**	−2,100	← Maximum negative cash flow, −$2,100
5	−500	1,200	**700**	−1,400	
6		1,200	**1,200**	−200	
7		1,200	**1,200**	1,000	← Cumulative net cash flow becomes positive during year 7.
8		1,200	**1,200**	2,200	
9		600	**600**	2,800	
10		600	**600**	3,400	
11		600	**600**	4,000	
12		600	**600**	4,600	
13		600	**600**	5,200	
Total	−3,800	9,000	**5,200**		

Actual value profit: +$5,200

The first column is our investment cash flow from Table 12. The income cash flow stream is from Table 14. The combination is shown under the heading of net cash flow (AFIT). Our goal in this chapter has been to determine this column. Note the negative numbers in the first five years. Our total net cash flow is $5.2 million, the difference between the income and investment cash flow streams.

Three important facts will be used from the cumulative cash flow shown in the last column. The *maximum* negative cash flow is one such fact. Even though the after tax investment totals $3.8 million for the School prospect, we are never out that much money. The School prospect earns money beginning in the fourth year and as such we only have out-of-pocket demands of $2.1 million.

The cumulative cash flow becomes positive in the seventh year. Thus between the sixth and seventh years, the income equals the after tax investment and a payout is achieved.

The third and final fact used from the last column is the total net cash flow itself. It is often called the actual value profit. In the following chapter, the reason for this terminology will become clear.

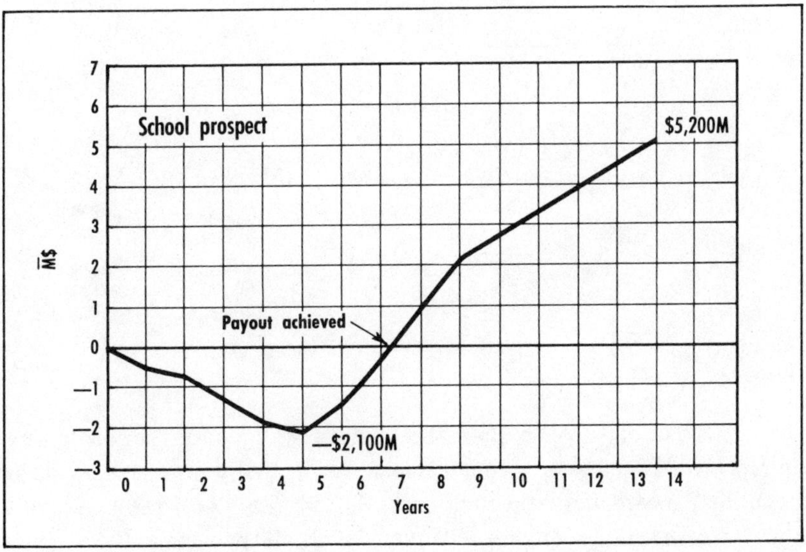

FIG. 5 CUMULATIVE NET CASH FLOW (AFIT)

On Fig. 5, these points are illustrated graphically. Note again the maximum negative cash flow, the total cumulative cash flow (actual value profit) and the year of payout.

Accounting versus cash flow How does cash flow analysis depict an investment differently from an accounting concept? The first fact to remember is that over the long haul the result is the same; i.e., the corporate net income (AFIT) is the same as the total net cash flow. There are differences, however, and they involve timing. The differences are illustrated in Table 16.

TABLE 16

Accounting concept vs. cash flow accounting concept

Period	*Invest.*	*Revenue*	*Deprec.*	*Other Exp.*	*Booked profit (before FIT)*
1	5,000	2,500	1,000	500	1,000
2		2,500	1,000	500	1,000
3		2,500	1,000	500	1,000
4		2,500	1,000	500	1,000
5		2,500	1,000	500	1,000
	5,000	12,500	5,000	2,500	5,000

Cash flow

Period	*Invest.*	*Revenue*	*Deprec.*	*Other Exp.*	*Booked profit (before FIT)*
1	5,000	2,500		500	−3,000
2		2,500		500	2,000
3		2,500		500	2,000
4		2,500		500	2,000
5		2,500		500	2,000
	5,000	12,500		2,500	5,000

The upper half of Table 16 illustrates on a before tax basis the accounting concept. Although the investment is made in the first year, it is reflected in accounting records only through depreciation; (i.e., the $5,000 does not enter the cash flow stream but is deducted at $1,000 per year as the revenue flows in from that investment). Note the uniform income or booked profit shown over the five-year period. Compare this illustration with the cash flow concept. In the cash flow concept, the $5,000 investment is placed in the cash flow stream in year one; it produces a negative income of −$3,000 compared to the positive $1,000 first year booked profit. Compare subsequent years. Note the income differences. At the conclusion of the investment, year 5, the total booked profit and the cash flow (BFIT) are the same. This sameness is the important point to remember.

Summarizing the relationship on Table 16, we note:

1. The accounting procedure matches booked outlays against revenue as it is produced. It illustrates the effect of an investment on the overall corporate books. Its results report earning power of the corporation to the stockholders.
2. The cash flow procedure matches outflows against inflows when the disbursement is made or cash received. Its primary use is to measure annual and cumulative cash flow and to provide a standard basis for comparing one investment with another. It places emphasis on year zero.
3. In both procedures the tax concepts are shown according to statutory requirements. As you will see in the next few paragraphs, the income tax generated by a successful investment is entered into each procedure in the same manner; and the taxes per year are identical.

To illustrate further the comparison between cash flow and accounting procedures, Table 17 was prepared. It makes the comparison for the School prospect. The first difference you will note is the timing illustrated in the first column. It shows the timing of cash outlays under accounting principles. Remember we stated that in the accounting procedure we related cash outlays to time as the income is produced. Thus the cash outlays to be reflected on the corporate books begin as revenue is produced and are reflected through depreciation and depletion. Column one is shown only for illustration purposes and does not enter the calculation of booked net income. Note that column four totals $3,600,000, the same as column one.

To calculate booked net income, we begin with the income cash flow stream before Federal income tax, the same as column seven, Table 14. The Federal income tax in column five includes investment tax credits which are preceded by a plus sign (+). These credits and expensed items affect booked net income until the fourth year. Inflows of cash start in the fourth year and depreciation of the investment begins. Note that the depreciation totals $3,600,000, the same as the capitalized (booked) investment shown on Table 11.

TABLE 17

Accounting concept vs. cash flow—M$
School prospect

Year	1 Cash outlay for booked inv.	2 Income cash flow before FIT	3 Exp. invest. items	4 Booked depln. & deprec.*	5 FIT	6 Booked net income (profit)	7 Net cash flow (AFIT)
0							−500
1	600		200		+100	−100	−200
2			1,200		+600	−600	−600
3	900		100		+400	300	−600
4	2,000	2,090	300	480	+10	1,320	−200
5	100	2,090	800	480	490	320	700
6		2,290		480	1,090	720	1,200
7		2,290		480	1,090	720	1,200
8		2,290		480	1,090	720	1,200
9		1,130		240	530	360	600
10		1,130		240	530	360	600
11		1,130		240	530	360	600
12		1,130		240	530	360	600
13		1,130		240	530	360	600
Total	3,600	16,700	2,600	3,600	5,300	5,200	5,200

*Modestly simplified for illustration purposes.

In column five, the FIT of $5.3 million is the algebraic sum of the $7.7 million and the $2.4 million in tax credits, Table 12. Column six, booked net income, is calculated by subtracting columns 3, 4, and 5 from column 2. The income cash flow less expensed items, depreciation and Federal income tax equals booked net income or corporate profit AFIT. In total ($5.2 million) it is the same as the net cash flow AFIT. The yearly difference results from the difference in timing of deductions caused by depreciation. Columns 6 and 7 allow a direct comparison between booked net income and net cash flow.

A final illustration of the accounting concept versus net cash flow is shown on Fig. 6. The total cumulative cash flow is the same as total corporate profit. The corporate books, however, show a much different plot on cumulative profit; e.g., the maximum negative profit is only —$700,000 compared to a maximum negative cash flow of —$2,100,000.

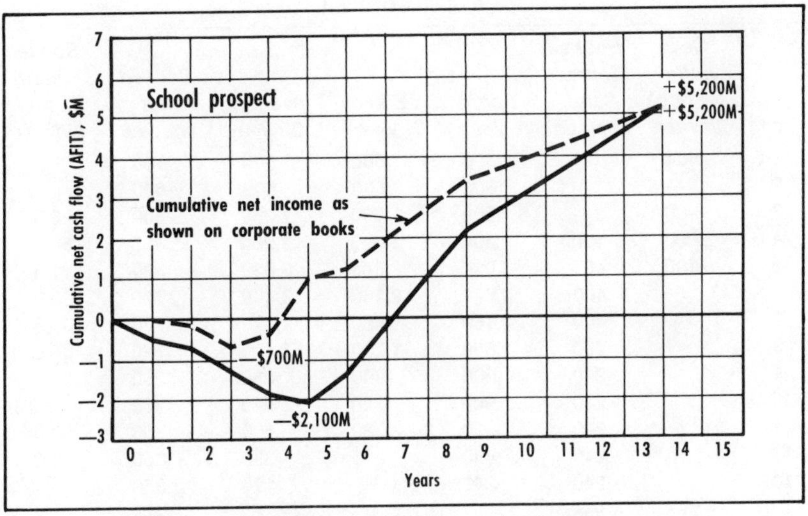

FIG. 6 ACCOUNTING CONCEPT VS. NET CASH FLOW

The comparison on Fig. 6 illustrates how accounting records reflect the return of cash versus cash flow analysis. The difference is not surprising, however, when the purpose of each is understood. The corporate books reflect the extinguishing of an investment coincident with its produced revenue. Accounting principles in all industries match depreciation to the revenues produced by the depreciating investment.

Cash flow analysis focuses on the available cash (or lack of it) caused by a specific investment. It is an evaluative tool. The seventh chapter will illustrate this concept. It is enough at this point to know that cash flow analysis is a useful tool for selecting the best investments when there are many investments from which to choose.

Booked return on investments The rate of return on booked net assets is a common financial yardstick. It is not often used in investment evaluation. In certain instances, however, it is desirable, particularly in a major investment, to know the effect on the corporate return on an investment opportunity. Table 18 illustrates the calculation of this yardstick for the School prospect.

TABLE 18

Booked return on investment
School prospect

Year	Cash outlay for book invest.	Booked deprec. and depletion	Beg. of year	End of year	Average for year	Booked net income	Booked return on invest.
1	600		500	600	550	−100	
2			600	600	600	−600	
3	900		600	1,500	1,050	300	28.6%
4	2,000	480	1,500	3,020	2,260	1,320	58.4
5	100	480	3,020	2,640	2,830	320	11.3
6		480	2,640	2,160	2,400	720	30.0
7		480	2,160	1,680	1,920	720	37.5
8		480	1,680	1,200	1,440	720	50.0
9		240	1,200	960	1,080	360	33.3
10		240	960	720	840	360	42.9
11		240	720	480	600	360	60.0
12		240	480	240	360	360	100.0
13		240	240		120	360	
Total	3,600	3,600			16,050	5,200	
		Avg. (for 13 years)			1,234.6	400.0	32.4%

The annual rate of return on booked investment is calculated by relating yearly booked net income (corporate profit) to the average booked investment. The average investment is calculated by dividing by two the sum of the beginning investment and the investment booked at the end of the year. Changes during the year—additions to investment and depreciation of existing investment—are reflected in the end of the year booked investment.

When used for investment evaluation the yardstick sought is the return over the life of the investment. It is calculated as follows:

$$\text{Booked rate of return} = \frac{\text{Cumulative booked net income} \times 100}{\text{Sum of average annual booked investments}}$$

For the School prospect, this calculation is:

$$\frac{5{,}200}{16{,}050} \times 100 = 32.4\%$$

REVIEW

Chapter 5 is one of the key building blocks in the series. Having digested the necessary terms and tax concepts, cash flow analysis can be introduced. Pertinent points from this chapter are:

1. A company grows based upon its investment opportunities and its skill in selecting the best investments from those available.
2. Cash flow analysis is a method of matching inflows and outflows. It is the first step in investment evaluation. The net cash flow is the algebraic sum of inflows and outflows.
3. Any investment is, in economic reality, a purchase of a series of future annual cash flows. Cash flow analysis is the technique by which these flows are determined.
4. The total net cash flow from an investment is exactly equal to the total net corporate profit. Annually, however, cash flow differs from corporate profit because of depreciation.
5. Three important factors are developed from cumulative net cash flow—namely, the payout period, the maximum negative cash flow and the total net cash flow.

A PROBLEM

At this point, you may want to test what you have learned. A brief problem follows; work it to see if you have adequately learned the terms and concepts to date. The problem emphasizes the investment flow stream and the net cash flow stream. In addition to the necessary data to work the problem, two handy work sheets are included to make your calculations easier. The solution is shown in Appendix B.

CHRONOLOGY OF SUCCESSFUL PROSPECT

Problem 1

Year 0: Good geological lead results in lease purchase.

Year 1: Geophysical work confirms prospect.

Year 2: Additional acreage leased.

Year 3: Wildcat test of prospect results in new field discovery.

Year 4: Development wells drilled; lease production facilities installed.

Year 5: Dry development well drilled. Field fully developed.

Year 15: Reserve depleted. Field abandoned.

REQUIRED INVESTMENT*

Item	Amount ($M)	Year Incurred
Lease bonus	200	0
Additional lease bonus	50	2
Geophysical work	50 (80% capitalized)	1
Lease rental (1st year)	10	1
Lease rental (2nd year)	10	2
Discovery wildcat	300 (70% intangible)	3
Development wells	1,000 (60% intangible)	4
Lease facilities	200	4
Dry hole	200†	5

*Geological, scouting, overhead, etc. expenses not to be considered in this problem.
†Is it necessary to know per cent intangible for this item?

SUCCESSFUL PROSPECT
NET CASH FLOW

Year	*AFIT investment*	*Net income*	*Net cash flow*	*Cumulative net cash flow*
	$M	*$M*	*$M*	
0		—		
1		—		
2		—		
3		—		
4		300		
5		500		
6		500		
7		500		
8		500		
9		500		
10		500		
11		500		
12		500		
13		300		
14		300		
15		100		

Maximum negative cash flow in year ________.
Cumulative cash flow becomes positive during year ________.
Payout in year ________.
Total after-tax investment: $M________.
Actual value profit: $M________.

INVESTMENT CASH FLOW STREAM

Year	Capitalized	Tangible	Intangible	Expensed	Total Col. 3 & 4	FIT Credit	AFIT Investment	Year	Remarks
	(1)	(2)	(3)	(4)	(5)	(6)	(7)		
0								0	
1								1	
2								2	
3								3	
4								4	
5								5	

6 *Tomorrow's Dollars Today (Present Value Concepts)*

The concept of present value is one of the most valuable tools in investment evaluation, but definitely not the only tool. It is not a cure-all, but must be used with a thorough understanding of its advantages and limitations. Present value concepts must also be used in conjunction with other economic concepts for meaningful comparisons of investment opportunities.

You have heard the terms "discounted cash flow," "rate of return," and "present value profit." If you haven't previously been exposed to the basic principles of present value, these terms may seem shrouded by an aura of mystery. However, there is nothing new, magical or mysterious about present value economics. In fact the principles we shall discuss have been used by bankers and insurance men since "dealing in dollars" became a business. So, before we talk about present value principles, applied to the oil industry, let's talk a bit about the money business.

Interest What makes the money business work is interest. The ability of invested funds to earn interest makes the money business a profitable one. The exact amount of interest earned is a function of two things:

1. The interest rate (i.e., 4½%, 6%, 8½%, etc.).
2. The frequency of compounding (i.e., annually, quarterly, daily, etc.).

Throughout this chapter, for simplicity's sake, we will be talking only about interest compounded annually. Although more frequent compounding does make a difference, it does not significantly affect exploratory investment analysis. See Appendix C for some comments on compounding at faster than annual rates.

Future value of invested dollars Let's now look at how interest acts to increase the value of money over a period of time. Suppose we have a dollar to invest; furthermore we know where it can be invested at 10% compounded annually. We will invest our dollar now and call this point in time—"year zero." After one year (year one), our money is worth:

$$(\$1.00) + (0.10)\ (\$1.00) = \$1.10 \qquad \text{Equation 1}$$

Our dollar has earned 10 cents during its first year of investment. We can state Equation 1 another way by factoring out the ($1.00). It then becomes:

$$(\$1.00)\ (1 + 0.10) = \$1.10 \qquad \text{Equation 2}$$

In the second year, we earn interest on our original investment of $1.00 and also on the $0.10 interest earned in year one. At the end of two years, we have:

$$(\$1.10) + (0.10)\ (\$1.10) = \$1.21 \text{ or restated} \qquad \text{Equation 3}$$

$$(\$1.10)\ (1 + 0.10) = (1.10)^2 = \$1.21 \qquad \text{Equation 4}$$

In like manner, if our dollar remains invested for 10 years at 10% interest compounded annually, we would have $2.59 or 2.59 times our original investment.

The steps just enumerated are shown in Table 19. Several pertinent facts come from Table 19.

1. During the seventh year, our original investment doubles.
2. The equation for compounding our investment utilizes the number of the year as an exponent.

From the illustration, we can show how the growth of any investment at any interest rate can be determined:

If: i = interest rate
n = number of years at interest i compounded annually
I = the investment

Then: The future value of today's investment =

$$(\text{Investment})\ (1 + i)^n \text{ or}$$
$$FV = I\ (1 + i)^n \qquad \text{Equation 5}$$

Equation 5 can be used to calculate the future value of an investment of any size. Problem 1 at the end of the chapter will test your understanding of this point.

The present value of future dollars Let's look at the money business from another angle. Suppose we don't want to invest a whole dollar at 10%—but some amount less. We want our money to grow, through interest, to be worth one dollar at the end of one year. We ask ourselves a different question. How much must we invest now to have one dollar at year end? Mathematically we are asking:

$$\text{What number} \times 1.10 = \$1.00?$$

The number we are looking for is our investment, so the equation becomes:

$$I \times (1.10) = \$1.00$$

or

$$I = \frac{\$1.00}{1.10} = \$0.909 \qquad \text{Equation 6}$$

TABLE 19

Future value of $1.00 invested at 10%

Years hence	*Amount of 1 (Invested @ 10% rate)*
0	1.00
1	$(1.00) + (1.00)(.10) = (1.00)(1 + .10) = (1.00)(1.10) = 1.10$
2	$(1.10)(1.10) = (1.10)^2 = 1.21$
3	$(1.21)(1.10) = (1.10)^3 = 1.33$
4	$(1.10)^4 = 1.46$
5	1.61
6	1.77
7	1.95
8	2.14
9	2.36
10	$2.59 = (1.10)^{10} = (1 + .10)^{10}$ . . . etc.

Equation 6 tells us—if we invest 90.9¢ for one year at 10%, we will have a dollar at year end. Stating this thought another way—today's value of next year's dollar (at 10%) is \$0.909. A dollar of income received next year is worth only 90.9¢ today because of the loss of one year's interest at 10%.

Next let's consider the same problem, only now we want our investment to grow to a dollar in two years.

$$I\,(1 + 0.10)\,(1 + 0.10) = 1.00$$

$$I\,(1.10)^2 = I\,(1.21) = 1.00$$

$$I = \frac{1.00}{1.21} = 0.826 \qquad \text{Equation 7}$$

Equation 7 tells us that if we can earn 10% compounded annually, we need only invest \$0.826 to have one dollar in two years.

In investment analysis the today's value of future dollars is called the present value; and Equation 7 is usually written:

$$PV = \frac{1.00}{1.21} = 0.826$$

It would be read—that the present value of one, discounted at 10% for two years, is 0.826. Again, this concept can be illustrated by a table:

TABLE 20

Present value of 1, discounted at 10%

Years hence	*Present value of 1 (Discounted @ 10% rate)*
0	1.000
1	$(1.00) \div (1 + .10) = (1.00) \div (1.10) = 0.909$
2	$(1.00) \div (1.10)^2 = (1.00) \div (1.21) = 0.826$
3	$(1.00) \div (1.10)^3 = (1.00) \div (1.33) = 0.751$
4	0.683
5	0.621
6	0.564
7	0.513
8	0.467
9	0.424
10	0.386 . . . etc.

TABLE 21

Present value of 1

$$\frac{1}{(1+i)^n}$$

	Discount rate										
Years hence	*1%*	*3%*	*5%*	*6%*	*7%*	*10%*	*15%*	*20%*	*30%*	*40%*	*50%*
1	.990	.971	.952	.943	.935	.909	.870	.833	.769	.714	.667
2	.980	.943	.907	.890	.873	.826	.756	.694	.592	.510	.444
3	.971	.915	.864	.840	.816	.751	.658	.579	.455	.364	.296
4	.961	.888	.823	.792	.763	.683	.572	.482	.350	.260	.197
5	.951	.863	.784	.747	.713	.621	.497	.402	.269	.186	.132
6	.942	.837	.746	.705	.666	.564	.432	.335	.207	.133	.088
7	.933	.813	.711	.665	.623	.513	.376	.279	.159	.095	.059
8	.923	.789	.677	.627	.582	.467	.327	.233	.123	.068	.039
9	.914	.766	.645	.592	.544	.424	.284	.194	.094	.048	.026
10	.905	.744	.614	.558	.508	.386	.247	.162	.073	.035	.017
11	.896	.722	.585	.527	.475	.350	.215	.135	.056	.025	.012
12	.887	.701	.557	.497	.444	.319	.187	.112	.043	.018	.008
13	.879	.681	.530	.469	.415	.290	.163	.093	.033	.013	.005
14	.870	.661	.505	.442	.388	.263	.141	.078	.025	.009	.003
15	.861	.642	.481	.417	.362	.239	.123	.065	.020	.006	.002
16	.853	.623	.458	.394	.339	.218	.107	.054	.015	.005	.002
17	.844	.605	.436	.371	.317	.198	.093	.045	.012	.003	.001
18	.836	.587	.416	.350	.296	.180	.081	.038	.009	.002	.001
19	.828	.570	.396	.331	.277	.164	.070	.031	.007	.002	
20	.820	.554	.377	.312	.258	.149	.061	.026	.005	.001	
21	.811	.538	.359	.294	.242	.135	.053	.022	.004	.001	
22	.803	.522	.342	.278	.226	.123	.046	.018	.003	.001	
23	.795	.507	.326	.262	.211	.112	.040	.015	.002		
24	.788	.492	.310	.247	.197	.102	.035	.013	.002		
25	.780	.478	.295	.233	.184	.092	.030	.010	.001		
26	.772	.464	.281	.220	.172	.084	.026	.009	.001		
27	.764	.450	.268	.207	.161	.076	.023	.007	.001		
28	.757	.437	.255	.196	.150	.069	.020	.006	.001		
29	.749	.424	.243	.185	.141	.063	.017	.005			
30	.742	.412	.231	.174	.131	.057	.015	.004			
31	.735	.400	.220	.164	.123	.052	.013	.004			
32	.727	.388	.210	.155	.115	.047	.011	.003			
33	.720	.377	.200	.146	.107	.043	.010	.002			
34	.713	.366	.190	.138	.100	.039	.009	.002			
35	.706	.355	.181	.130	.094	.036	.008	.002			
36	.699	.345	.173	.123	.088	.032	.007	.001			
37	.692	.335	.164	.116	.082	.029	.006	.001			
38	.685	.325	.157	.109	.076	.027	.005	.001			
39	.678	.316	.149	.103	.071	.024	.004	.001			
40	.672	.307	.142	.097	.067	.022	.004	.001			
41	.665	.298	.135	.092	.062	.020	.003	.001			
42	.658	.289	.129	.087	.058	.018	.003				
43	.652	.281	.123	.082	.055	.017	.002				
44	.645	.272	.117	.077	.051	.015	.002				
45	.639	.265	.111	.073	.048	.014	.002				
46	.633	.257	.106	.069	.044	.012	.002				
47	.626	.249	.101	.065	.042	.011	.001				
48	.620	.242	.096	.061	.039	.010	.001				
49	.614	.235	.092	.058	.036	.009	.001				
50	.608	.228	.087	.054	.034	.009	.001				

An equation also expresses the present value concept. It develops in a way similar to Equation 5. Using the 10th year as an illustration:

The present value of 1, discounted at 10% for 10 years $= \dfrac{1}{(1 + 0.10)^{10}} = \dfrac{1}{2.59} = 0.386$

So if: i = discount rate

n = number of years discounted at rate i

Then $$\text{PV of } 1 = \frac{1}{(1 + i)^n} \qquad \text{Equation 8}$$

The message from Table 20 is—if you can invest your money today at any interest rate, then tomorrow's dollars are worth *less* than a dollar today. The table shows the discount factors for the 10% discount rate. Note how rapidly the value of future dollars diminishes. By the 10th year they are worth (today) only $0.386. Note also in year zero, Table 20, the 1.000 figure. This number symbolizes the fact that a dollar in the pocket today is worth *one* dollar—not more or less.

On Table 21 the discount factors are shown for eleven different discount rates, ranging from one to 50%. You will recognize the first few factors shown under the 10% discount rate.

Discounting a cash flow It was necessary to go through the preceding steps to enable us to comprehend discounting. One of the steps taken in investment evaluation is to discount a cash flow stream. When discounting a flow of cash to present value, we make two assumptions:

1. In making an investment we are, in effect, purchasing a series of annual future incomes.
2. Future incomes can in turn be reinvested in other income generating activities.

These two assumptions are important. Through reinvestment any income received early has more value than later income. If the discount rate is said to be the interest rate that can be earned through reinvestment, we can use discount factors to compare values of income dollars—at various points in time.

Early in the life of an investment (remember the School prospect), the annual cash flows are most often negative. In discounting a cash flow stream, the negative values are discounted also. Why? Because we assume that until investment funds are required, that money is invested elsewhere at a rate equal to the discount rate.

Perhaps by now you are catching a glimpse of the importance of the discount rate and the timing of investments. We are about to clarify the concept of the time value of money. Let's re-examine Table 21. We will take a few numbers from this table and construct a new one to illustrate the significance of various discount rates—with time.

TABLE 22

Comparing factors from "present value of 1"

Years hence	*Actual value (undiscounted)*	*Discount rate*		
		5%	*10%*	*20%*
0	1.000	1.000	1.000	1.000
1	1.000	.952	.909	.833
5	1.000	.784	.621	.402
10	1.000	.614	.386	.162
20	1.000	.377	.149	.026

A comparison of present value is made for the zero year and years 5, 10, and 20. Note the undiscounted value. It will henceforth be called the actual value. In Table 22, it is always 1.000 and it is always 1.000 for any discount rate in the zero year.

Check the 10-year line. At 5% money ten years hence is worth only 0.614 per dollar. Compare the 20% rate. In the 10th year a dollar is only worth $0.162 if you could have invested it at 20%. In the 20th year at a 20% discount rate, a dollar has a value of only $0.026.

The effect of discounting is shown graphically on Fig. 7. Note first how fast the discount factor decreases at high rates, such as 50%. To compare the effect of various rates on a cash flow stream, start at the left of Fig. 7 on the 0.1 discount factor line. (Think of this factor as a dollar worth 10 cents.) Moving to the right, check the intersection with each discount rate curve. At 50% by the sixth year income flow is discounted below 10 cents. At 30% the 10 cent line is reached in 9 years, at 20% in 13 years. These curves demonstrate an important fact. Any event beyond the 10th year in a cash flow stream (inflow or outflow) has little effect on present value profit at rates higher than 20%. Even at a 10% discount rate, the significance of investments or revenues is sharply reduced beyond 10 years.

Now, to discount a cash flow; again, we shall use the School prospect. In Table 23 you will recognize its cash flow stream and in column 2 the 10% discount factors. These factors are multiplied times the numbers in column 1 to get the present value of the cash flow discounted at 10%. Note that no discounting takes place in year zero. Today's dollar is worth a dollar. As mentioned earlier both negative and positive values are discounted. The total of column 3 is the present value of the entire net cash flow stream, discounted at 10%. It totals $1,583,000 compared to the actual value of $5,200,000.

Column 3 tells us—if we have alternate investments at 10%, then the present value of future incomes from the School prospect is worth $1,583,000; or the School prospect will provide a 10% return on its cash flow stream plus an additional $1,583,000.

In column 4 the 20% discount factors are shown. If they are multiplied times the net cash flow (column 1) of the School prospect, column 5 results. It is the present value of the net cash flow stream discounted at 20%. Now the present value for the 13 years of annual incomes totals only $189,000. If we have alternate investments at 20% then the School prospect will return 20% plus $189,000.

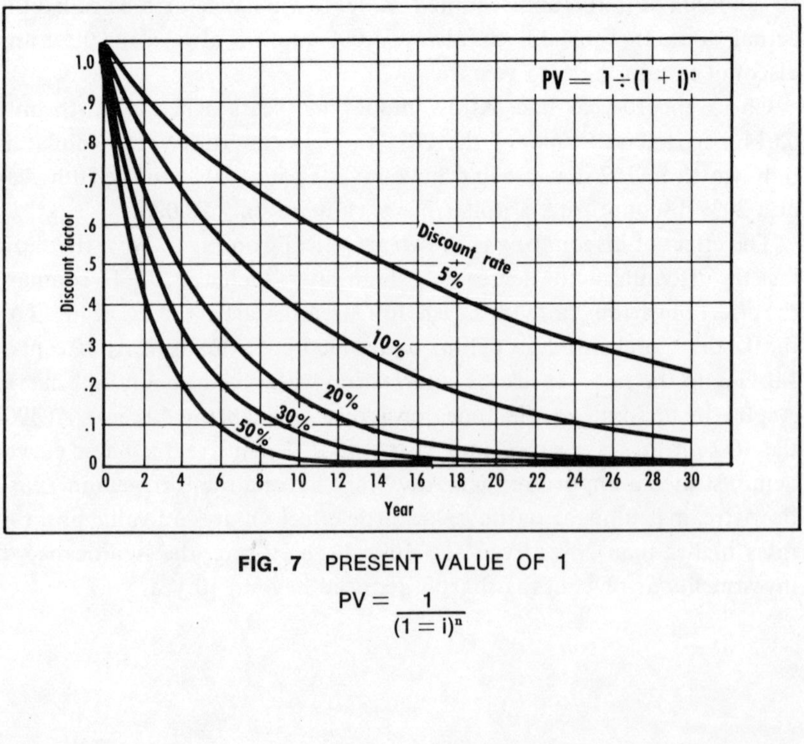

FIG. 7 PRESENT VALUE OF 1

$$PV = \frac{1}{(1 = i)^n}$$

You can already envision one of the uses of this tool. If you set a target rate of return then two cash flow streams can be discounted at that rate. The one having the largest present value is the better investment (based only on this one yardstick).

Constructing a present value profile We are now ready for another quantum leap in our investment knowledge. We have proceeded far enough to be able to construct a present value profile. It is one of the most useful of all evaluating tools.

Our example will be the School prospect. The present value profile for the School prospect is shown on Fig. 8. On such a chart, the present value is plotted for various discount rates. The discount rate is usually the horizontal axis—the present value, the vertical axis.

Note the 0% discount point in Fig. 8. You will recognize the actual value profit (AVP) of $5,200,000. Remember, this number is the undiscounted

TABLE 23
Discounting a net cash flow stream
School prospect

Year	*1 Net cash flow (after FIT) M$*	*2 Discount factors 10% rate*	*3 Present value discounted @ 10%*	*4 Discount factors 20% rate*	*5 Present value discounted @ 20%*
0	−500	1.000	−500	1.000	−500
1	−200	.909	−182	.833	−167
2	−600	.826	−496	.694	−416
3	−600	.751	−451	.579	−347
4	−200	.683	−137	.482	−96
5	700	.621	435	.402	281
6	1,200	.564	677	.335	402
7	1,200	.513	616	.279	335
8	1,200	.467	560	.233	280
9	600	.424	254	.194	116
10	600	.386	232	.162	97
11	600	.350	210	.135	81
12	600	.319	191	.112	67
13	600	.290	174	.093	56
Totals	$5,200 M		$1,583 M		$189 M
	(Actual value profit)		*(Present value profit discounted @ 10%)*		*(Present value profit discounted @ 20%)*

profit. Move to the 10% discount rate. Here you will recognize the present value profit (PVP) of $1,583,000 originally shown as the total of column 3 on Table 23. Next check the PVP at 20%. Again this point is from Table 23. The PVP is $189,000 discounted at 20%.

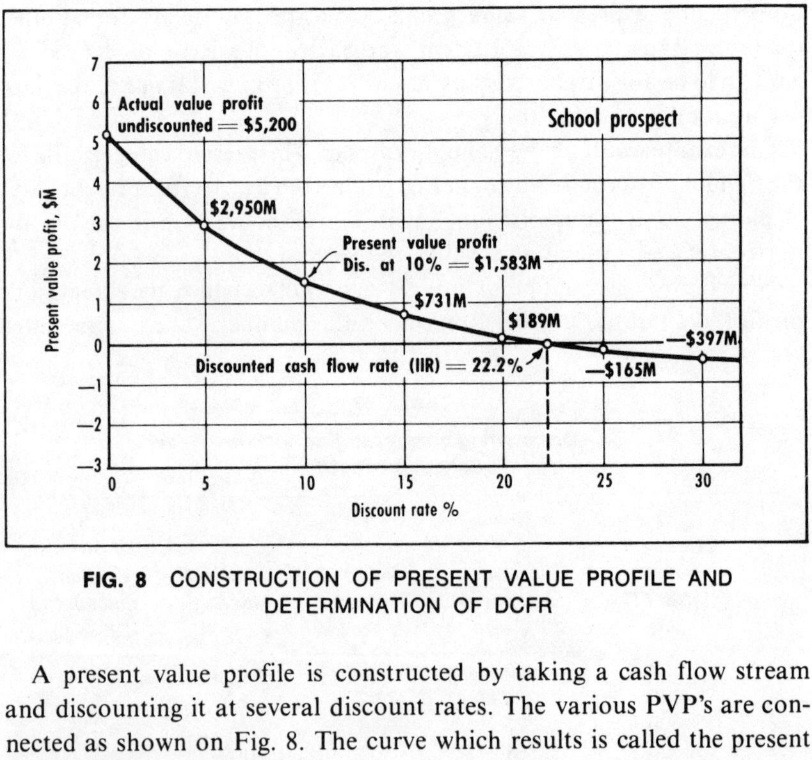

FIG. 8 CONSTRUCTION OF PRESENT VALUE PROFILE AND DETERMINATION OF DCFR

A present value profile is constructed by taking a cash flow stream and discounting it at several discount rates. The various PVP's are connected as shown on Fig. 8. The curve which results is called the present value profile.

Two other points can be observed from Fig. 8. Note where the profile crosses the zero PVP line. At this point the discount rate is large enough that it discounts the net cash flow stream to zero; i.e., the sum of the discounted negative and positive annual incomes is zero. This point is called the discounted cash flow rate (DCFR) or investor's interest rate (IIR). Note also that the profile goes below the zero present value line. For discount rates larger than 22.2%, the resulting PVP's are negative. What this tells us is—if you can invest money at amounts greater than 22.2% you would (in most instances) be better off to take such investments rather than the School prospect. However, as we shall see in the next chapter, there could be investments with a DCFR of 25% which are not necessarily better than the School prospect.

Review of figure 8

1. A present value profile is a plot of the PVP's resulting from discounting a net cash flow stream at various rates.
2. Three important features of the profile help us in investment evaluation. They are:
   a. The magnitudes of the actual value profit and present value profits at commonly quoted discount rates.
   b. The shape of the profile.
   c. The discounted cash flow rate (DCFR).
3. The discounted cash flow rate is that rate which causes a net cash flow stream to be discounted to zero. The DCFR may also be thought of as the rate which causes the sum of the discounted outflows and inflows to equal the net cash outlay in year zero. (Remember year zero is undiscounted.)

Calculating a DCFR The discounted cash flow rate must be calculated by trial and error. It can also be determined graphically as shown in Fig. 8.

In most modern corporations, computer programs are available to calculate the DCFR. You may occasionally be faced with the need to calculate it manually with a desk calculator, adding machine, or slide rule. Where do you start? What discount rate should you first apply to the net cash flow stream? A very useful rule of thumb aids us here.

The reciprocal of payout in years times 100 is often close to the DCFR, especially for projects with equal annual cash flows. For example, if the cumulative cash flow shows payout in the fifth year, try 20% as your first discount rate:

$$\frac{1}{\text{Payout}} \times 100$$

$$\frac{1}{5} \times 100 = 20$$

If the PV at 20% is *negative,* you have used too large a discount. Try a smaller discount rate; do this until you have a positive PV. Three points (discount rates) are usually enough to determine the DCFR graphically.

Plot the three points as in Fig. 8 and where the curve crosses the zero line of PV, read the DCFR.

Another interesting and useful rule of thumb is the "rule of 72." Dividing the number 72 by the years it takes to double your investment will give you the interest rate. Try it.

Another method of viewing the DCFR One can also view the discounted cash flow rate as the *maximum* interest rate which could be paid on borrowed capital and still break even.

Consider the following example. We have said that an investment yielding a 20% DCFR would allow us to borrow money at that rate and break even. This concept is illustrated in Table 24.

In column 1 the after-tax net cash flow shown in column 2 is discounted at 20%. The sum of the annual discounted values is zero, indicating a DCFR of 20%. Note in year zero that we borrow $3,000 which is undiscounted. Switch your attention now to the interrelationship between columns 2, 3, 4, and 5.

Our investment starts off with a negative cash flow of $3,000. In year one it earns $1,000. We pay $600 interest (column 3), apply $400 on the principal (column 4) and have an unpaid balance of $2,600 (column 5). In year two, our cash flow is again $1,000. Again we pay interest at 20% and apply some money to the principal. Our unpaid balance goes down each year, finally to zero in year six. Our cash flow was sufficient to pay off the $3,000 note and also to pay interest on the unpaid balance at the rate of 20%. The amount of interest paid was $2,030, equal to our after-tax net cash flow. We are making the assumption in this illustration that tax credits from interest paid have been taken into account in column 2, the net cash flow.

If we had paid a higher rate of interest, our cash flow would have been exceeded. A lower rate of interest would have yielded extra cash flow; so the DCFR represents the maximum interest we could pay on borrowed money and still break even on the investment.

The reinvestment controversy The DCFR results from trial and error discounting of a net cash flow stream. It is often referred to as the internal rate of return because the rate is "internal" to the project. For this reason you will find references in the literature which state that the reinvestment of future cash flows does not need to be considered when using DCFR.

TABLE 24

The concept of a discounted cash flow rate

Year	1 *Present value disc. @ 20%*	2 *After-tax net cash flow*	3 *Interest @ 20% on unpaid balance*	4 *Payment to reduce principal*	5 *Unpaid balance*
0	−3,000	−3,000	—	—	−3,000
1	833	1,000	−600	400	−2,600
2	694	1,000	−520	480	−2,120
3	579	1,000	−424	576	−1,544
4	482	1,000	−309	691	−853
5	402	1,000	−172	828	−25
6	10	30	−5	25	−0−
	−0−	2,030	−2,030	3,000	
(20% DCFR)					

However, such is not the case. Granted the discounting to zero is internal to the investment—but once you use DCFR to compare to another investment you have automatically made the reinvestment assumption. Pollack[4] (1961) showed convincingly that in comparing investments the reinvestment assumption *is* made.

Why is this controversy important? It is important because of the loss of validity of very high DCFR's. If your company investments average 12% return then DCFR's of 60 to 80% are unrealistic in that you do not have multiple opportunities to reinvest at such high rates.

Nevertheless the controversy does not take away the value of a high DCFR as reflecting a quick payout and a fast return of your investment. We shall see in the next chapter some recent attempts to develop yardsticks which compensate for this flaw of the DCFR.

Comparison of DCFR and booked return Comparisons are often made between "rates of return" calculated by various means. In an accounting sense, rates of return are returns related to net assets or stockholders' equity, etc.

The DCFR is not a rate of return in that sense. It emphasizes the timing of inflows and outflows to a greater extent than does a return on net assets.

For most evaluations the booked rate of return will be larger than the DCFR. For the School prospect:

- the booked return was 32.4%
- the DCFR was 22.2%

As development costs increase, the two rates tend to converge near the same value. There are numerous references in the literature to other "rates of return" and evaluation methods.[1,2,3] New concepts in evaluation should be expected to emerge as new needs appear and better methods are discovered to describe investment opportunities. As will be shown in the next chapter, evaluation techniques and yardsticks are still evolving.

The best way to compare the DCFR and a return on net assets is to note their intent. A return on assets measures the profit over a period as related to the capitalized investments. It is a standard business method common to all profit making organizations. The DCFR is not a return on assets. Its intent is to help in the comparison between investments. It measures their cash flow relative to year zero on a common basis, not their assets. These differences allow a generalization about the two terms. The DCFR is for evaluation purposes to help (with other yardsticks) choose the best investment. A return on assets reports, in standard accounting concepts, the money earned by a firm's assets. The two terms are distinctly individual in nature and intent and are not comparable.

Present value profit (PVP) and inflation Although we shall have more to say about inflation, one aspect of discounting should be noted. If the minimum guideline for your investments is 10% (as illustrated in this book) it probably makes no provisions for inflation. Some analysts have raised the discount factor to include inflation, feeling this represents an adequate way of compensating for the loss in value of future dollars.

For example if inflation is expected at 5% per year, then discounting the cash flows at 15% would include 10% for the minimum guideline and 5% for inflation. This method demonstrates one way to handle inflation. There are other ways to be covered in a more complete handling of inflation in a later chapter.

REVIEW

Present value concepts are really time value concepts. The yardsticks available from present value relationships are reviewed in the next chapter. Nevertheless, we can already understand some of the importance of the time value of money.

Present value concepts are not difficult to use or understand if one remembers that money invested earns income called interest. In a profit making organization, one must remember that the investor can always leave his money in the bank, at almost no risk, and get a small return. Thus the profit making organization which has risks of failure must achieve a return higher than a bank; otherwise it cannot attract new capital for investment.

PROBLEMS

1. If you invest $1,000 at 10% interest, how many years elapse before you have one million dollars? Clue—remember how many years it takes to double your money at 10% compounded annually.
2. Using only a pencil and a columnar sheet, combine the investment and income cash flows which follow. Calculate the actual profit, payout and maximum negative cash flow. Then discount the annual cash flow stream using the discount factors given to get a DCFR.

Calculating a DCFR

	Cash flows—AFIT		*Rounded discount factors*			
Year	*Investment*	*Income*	*10%*	*15%*	*25%*	*30%*
0	−200	—	1.00	1.00	1.00	1.00
1	−900	200	.90	.90	.80	.75
2	−100	300	.82	.75	.65	.60
3		400	.75	.65	.50	.45
4		400	.67	.55	.40	.35
5		300	.62	.50	.33	.25
6		300	.55	.45	.30	.20
7		200	.50	.40	.24	.16
8		100	.47	.35	.18	.12
9		100	.42	.30	.14	.09

Answer: Appendix B.

BIBLIOGRAPHY

1. Kaitz, Melvin, "Percentage Gain on Investment—An Investment Decision Yardstick," Journal of Petroleum Technology, May 1967.
2. Arps, J. J., 1958, "Profitability Analysis of Proposed Capital Expenditures for Development Drilling and Appraisal of Producing Properties," AIME Petroleum Engineer Transactions, V. 213.
3. Hugo, G. R., "Farmouts Can Make Capital Efficient," Oilweek, October 2, 1967.
4. Pollack, Gerald A., "The Capital Budgeting Controversy: Present Value vs Discounted Cash Flow Methods," N.A.A. Bulletin, Nov. 1961, pg. 5.

7 Measuring an Opportunity

As we start the seventh chapter, let's review our progress to date; and we have made considerable progress.

1. We understand a few new terms needed for our investment evaluations.
2. We are aware of the significance of tax planning.
3. We can construct a cash flow stream.
4. We have a beginning understanding of the time value of money.

We now have sufficient knowledge to begin measuring our opportunities. We do this by comparing opportunities by means of investment yardsticks. This chapter will help us get acquainted with standard investment yardsticks.

Evolution of investment evaluation For a change of pace, we are going to indulge in a bit of history—to show how we got to where we are in investment evaluation, and why.

Throughout the history of the oil industry, there has been a gradual evolution in investment policies, opportunities, and evaluations. Yardsticks have been developed with time to fit the state of knowledge, skill, technology and even political conditions. When one views the changes to date, future changes seem to be implied.

A. Pre-proration era

In the pre-proration era, prior to 1930, the industry operated in an unstable economic climate. The instability was caused by numerous factors:

1. The ubiquitous cyclic nature of discoveries caused vast ranges in petroleum supplies from shortage to surplus.
2. Violent supply trends caused or abetted equally violent price trends.
3. Limited knowledge about fluid behavior in reservoirs, the geologic configuration of reservoirs, and the large numbers of individual operators conducting producing operations over bits and portions of an oil pool inhibited proper corrective measures.

Under such conditions of instability, the most important consideration for the producer was to get his money back quickly. How much it cost and the delay in return of capital were his prime yardsticks. Payout, or payback, became one of the first important yardsticks. Usually measured in years, payout is the time required for the opportunity to return the original investment. The shorter the time, the lower the risks.

B. The beginning of proration—1930-1942

The primary concern of the regulatory agencies in establishing uniform measures of control was conservation. However, one of the byproducts of sound conservation was an era of stable prices. Producers could plan a little further ahead. The consumer also gained. He had assurances of a continuous stable supply for his energy needs.

For the producer price stability meant the ability to forecast future income for periods longer than a few months. He could estimate total future profit with more certainty. He could even borrow a little money on a good producing property. For the first time, the total quantity of money to be gained could be estimated with reasonable accuracy and the ratio of this amount to the original investment now had meaning. The profit-to-investment ratio could be considered an important economic yardstick. Two investments with identical payouts could be meaningfully compared on another basis—profit-to-investment ratio.

Reservoir engineering was born during this period. It also was a pertinent factor in enabling the producer to forecast his future income with greater accuracy.

In the pre-proration era and during the proration era (pre-1943), the cost of finding hydrocarbons was nominal. Discoveries were shallow, acreage cheap and there were lots of unexplored basins. Although the seismograph was introduced and the first formal exploration programs planned, yardsticks were, in most companies, applied only to exploitation investments. It was the development well and its related facilities which received the most economic scrutiny.

C. The war years, 1943–1947

The war years produced a capacity production situation with sharply limited material supplies. Development of existing fields was given the highest priority and many attractive exploration investment opportunities lay dormant for lack of manpower and material. The scarcity of these two factors became more the limiting criteria than normal economic yardsticks. All activities had war time priorities. Such materials as steel, transportation equipment, cement, etc. were allocated based on government established priorities.

Little change in investment evaluation occurred during this period.

D. From capacity to surplus, 1948–1957

After World War II the pent-up demand for hydrocarbons soared. Oil operators, still suffering from manpower and material shortages, strained to meet demands.

In 1948 a capacity situation existed for most of the year. For a time even Texas produced at its maximum efficient rate (often called MER). Spurred by the needs and with price responses, the petroleum industry undertook an extensive exploration and exploitation program. The program was so successful that between 1948 and 1962 the state of Texas, in response to an ever increasing supply, reduced its market demand factor. This factor declined from 100% in 1948 to a low of about 26% in 1962.

There were other developments during this period. The rapid increase in demand brought upward changes in the price of crude oil and natural gas. The change was particularly significant for natural gas, long an

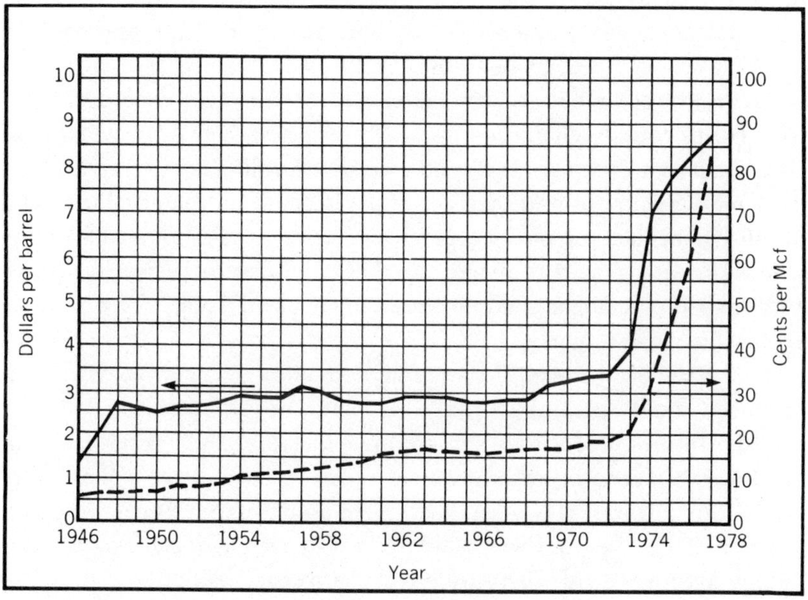

FIG. 9 AVERAGE U. S. WELLHEAD PRICE FOR CRUDE OIL AND NATURAL GAS

underpriced energy source. Natural gas cannot be easily stored; and the cost of transporting it is about three times the cost of transporting an equivalent amount of BTU's in the form of crude oil. However, it is the cleanest, pollutionwise, of all hydrocarbon fuels and at this time could be secured under long term contracts favorable to the purchaser.

The changing prices for both crude oil and natural gas are shown on Fig. 9. Natural gas prices have increased at a faster rate than oil prices, as this cheap fuel approaches its competitive BTU value. On a strictly BTU basis, it takes about 6 Mcf to equal 1 barrel of crude oil. Throughout most of the period shown on Fig. 9, you could get from 20 to 25 Mcf for the same price as a barrel of crude oil.

In the post embargo era average prices for crude oil in the United States have increased to about $8.50 per barrel—but are still below the world price. The average price as explained earlier is a combination of old oil (at the lowest price) and new oil (at a higher price). Even at today's prices the consumer can still buy BTU's in the form of natural gas cheaper than the equivalent BTU's in crude oil. Natural gas is still underpriced.

Much of the natural gas sold is under long term contract. Only new gas in the intrastate market, which sells for about $2.00 per Mcf, is approaching the BTU cost of an equivalent amount of oil.

As the price for natural gas increased, new reserves were developed from basins considered uneconomic at lower prices. The Arkoma basin of Oklahoma and Arkansas is a good example. Drilling is expensive in this hard rock country and known gas reserves could not be developed at a profit. About the time prices began to increase, the technology of drilling with air in hard rocks was introduced. The combination of increased price and air drilling made a previous unattractive basin a place for profitable exploration investments.

During the hectic days to supply an exploding demand for petroleum products, some operators forgot about interfuel competition. They assumed a future of continuous crude price increases. As a result many of the efficiencies later to be used were deferred by several years. They were introduced in the 1960's when costs were going up but prices were not.

The 1948–1957 period was characterized by rapidly increasing costs of finding hydrocarbons. There were few changes in evaluation criteria.

E. The period of surplus, 1957–1966

The best picture of surplus capacity is shown by Fig. 10. On this figure is plotted a history of Texas proration. The market demand factor (%) is plotted from 1946 through 1977. A steady decrease in market demand factor followed the capacity period of 1948. The sharpest drop came in 1958. The closing of the Suez Canal in 1957 had temporarily held demand at levels near those of the preceding years. After the Canal opened, the large surplus capacity in the U. S. forced Texas to prorate production to the lowest level in its history. The low point came in 1962 when the market demand factor was about 26%.

The industry has received much criticism, particularly from economists, because of its proration policies. The state of Texas has shared in this criticism. But Texas had several objectives in this matter and they were reasonable ones. The most important factors were conservation and protection of individual property rights. Each property owner was given his pro rata share of the market even though this market was restricted.

Another factor was revenue. The tax levied on petroleum and natural gas by the state is based on value. To see this value lowered by a temporary surplus would mean a permanent loss in revenue for Texas, something which the state leaders did not wish to happen. A few states have taxes not related to value. To protect its revenue for the citizens of Texas, to protect property rights, and to maintain proper conservation practices, severe proration was instituted. The spare capacity thus created was to be of considerable importance in the second closing of the Suez Canal.

The economic effect of severe proration is to limit production and revenue, lengthen payout and sharply restrict a company's ability to grow. We noted the reasons Texas was particularly hard hit by surplus capacity. Louisiana, whose tax is related to the barrel not value, also restricted production during this period. Again, the primary concern was conservation—to avoid physical and thus economic waste.

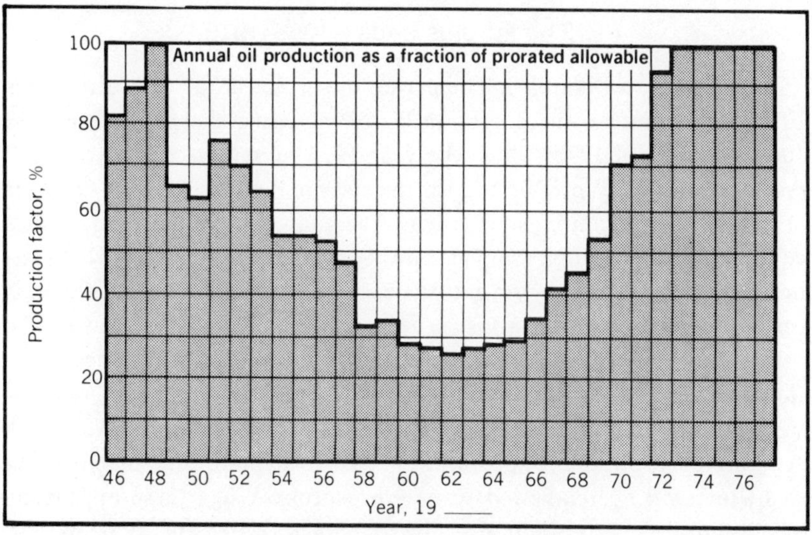

FIG. 10 HISTORY OF TEXAS PRORATION 1946–1977

Severe proration resulted in a close scrutiny of all investments. Particular emphasis was placed on those investments which were capable of a high cash flow. The result was a growing appreciation of yardsticks associated with concepts of the time value of money. The present value profile, DCFR, and even accelerated income investments were considered. An accelerated income investment is one made primarily to obtain an advantage in timing of income. Often it implies for an oil property no increase in ultimate recovery, but greater recovery in the early years with reduced income in later years. Such investments can give a present value profit, but an actual value loss; e.g., going from 80 to 40-acre spacing would be an example of an accelerated income investment. If there were no increase in recovery, the extra investment would mean an actual value loss compared to 80-acre spacing—even though larger revenues early in the cash flow stream would yield a PVP.

As companies looked for larger cash flows, and optimum selections of investments and reviewed new concepts of risk evaluation, the resultant evolution in yardsticks was considerable. In addition, the advent of the computer significantly changed the number of alternates which could be reviewed in a short time.

F. The surplus ends—1966-1970

Refer again to Figure 10. The period of surplus crude ends after 1966. From that point on the proration factor increased sharply and reached 100% in April of 1972. Thus, the year 1973 represents the first year of essentially capacity production for the state of Texas. The state has been at capacity production since then. There are individual fields which could produce at a slightly higher rate. These fields produce at MER rates deemed by the RRC to prevent waste. Excessive producing rates can, in certain reservoirs, result in lost production.

The trend in the market demand factor indicates the enhanced cash flow for the industry in the 1970's. This factor plus increased prices at the wellhead provides more funds which seek opportunities for reinvestment. As long as profitable opportunities exist in exploration, much of the cash flow will be reinvested in new prospects. When these opportunities diminish the funds will seek investments in related energy fields. The purchase of coal and uranium reserves by a few oil companies indicates possible avenues for diversification. However, the sharp increases in rig use and seismic crews indicate a good climate for investments in the petroleum industry. Inflation and over-regulation could take the bloom from this surge of activity if they are not held in check.

G. The era of regulation—after 1970

The full impact of bludgeoning regulation on yardstick evolution is unknown. The increased necessity of forecasting future prices, delay, permitting, etc. has stretched the abilities of all analysts. The regulatory era has had the same economic impact as severely restricted production—because in the end it results in lost income or delayed income. As a result payouts lengthen, revenue is uncertain, and start up dates become only approximate guesses unrelated to technical ability. The methods by which analysts are coping with this new era are evolving. In numerous instances probabilities of occurrence are being assigned to factors once considered fixed.

The present result of the regulatory era is investment evaluations and forecasts with a significantly greater degree of uncertainty. Even forecasts of income from existing fields take on new uncertainties as rules and regulations change. Continued uncertainty does not bode well as an investment climate.

Nevertheless, yardsticks to review investment opportunities will be developed to cope with new situations. History at least tells us this much.

Investment yardsticks Now back to measuring an opportunity. First let's review the term investment. In investment evaluation any expenditure made in the expectation of an increased profit can be regarded as an investment. (We usually look at the after-tax expenditures as we saw in Chapter 5.) Contrast this concept with the accounting definition where the investment refers only to those expenditures which are capitalized. Therefore, investment evaluation implies:

1. The expectation of future profits, usually involving both uncertainty and risk.
2. Income generated over a period of time.
3. A freedom of choice among investments, i.e., the discretion to select the best from various opportunities.

Investment yardsticks are the various criteria we use to help us in measuring, comparing and describing investment opportunities. In this chapter we will define most of the commonly used yardsticks and illustrate the use of yardsticks with example cases.

The ideal yardstick It would be extremely useful if there were a single yardstick which properly ranked each investment opportunity. We would probably call this measurement the ideal yardstick. It would have certain characteristics.

1. It would illustrate the effect on corporate profits of making a specific investment.
2. It would isolate only those investment opportunities which are acceptable to established guidelines.
3. It would always make the proper choice from a group of mutually exclusive opportunities. (In a group of investment opportunities, if only one may finally be selected, the investments are said to be mutually exclusive.)

Unfortunately, no ideal yardstick exists. As a result, several yardsticks are necessary for proper evaluation of an opportunity. Wise decisions are more likely when measuring an opportunity from several viewpoints. Some yardsticks require special care. They may be constructed differently by different individuals. When using any yardstick, be sure you understand how it was calculated. Was it before tax or after tax; booked investment or after tax expenditures; present or actual value?

A few opportunities are good investments almost by inspection; but very large investments, marginal investments, or high risk investments demand more careful scrutiny. Investments in areas unfamiliar to management (new geology or new industry) also require more evaluation. It is the bane of the investment analyst that his requester usually wants one more yardstick than those calculated—and that he spends a disproportionate amount of time evaluating marginal investments.

Some common investment yardsticks

A. *Yardsticks from cash flow analysis*

From Chapter 5 we learned that several useful factors are developed from cash flow analysis. Cash flow techniques provide directly many common yardsticks and are the basis for others. The yardsticks from cash flow analysis are:

1. The investment—both before and after tax. It may be separated into exploratory investments and producing investments. These are sometimes expressed on a unit basis (i.e., \$/bbl.). If so, the before tax expenditure is used.
2. Maximum negative cash flow. The largest sum of money, out-of-pocket at any one time, is called the maximum negative cash flow. You calculated this in the problem at the end of Chapter 5.

3. Ultimate net profit or loss. The cumulative net cash flow from a project is its ultimate net income (or profit). It is the sum of inflows minus outflows and is often called the actual value profit (AVP).
4. Ultimate net income ratio. The total actual value profit divided by the cumulative maximum negative cash flow is called the ultimate net income ratio.
5. Profit to investment ratio. The total actual value profit divided by investment is the profit to investment ratio. Please note that the term investment has not been defined. Be sure when you see a profit to investment ratio (P:I or P/I) that you understand just what investment is being used. Is it before tax or after tax? Originally this yardstick was used only with the capitalized investment. Use this yardstick with care and always define your investment.

B. *Yardsticks that reflect timing*

Certain yardsticks are related to time. Some have a calendar significance and others reflect the time value of money.

1. Project life. The length of a project in years is the project life. This yardstick has significance for a corporation. Its importance is related to a corporation's view of the future. Inflation, changing taxes, variable risks, changing energy patterns, etc. can encourage a company to favor either long or short project life depending on its goals.
2. Payout period. The time in years to return the after tax investment is called payout. The payout is calculated from the net cash flow stream. The point at which the cumulative net cash flow becomes positive is the project's payout.
3. Present value profit. A discounted net cash flow stream is called a present value profit—or if negative, a present value loss. It must always be expressed with an accompanying percentage which tells you the discount factor used to reduce the cash flow. Present value profits are usually shown at commonly quoted discount rates (10% for example).
4. Present value profile. The curve resulting from plotting present value profits versus their discount rates is called the present value profile. We shall review this yardstick in much more detail. Its configuration is important in investment evaluation.[1]

5. Discounted cash flow rate. (Sometimes called investor's interest rate.) The discount per cent which reduces a cash flow stream to zero is called the discounted cash flow rate.

C. *Yardsticks based on accounting concepts*

1. Booked investment. Those items which would be capitalized on the corporate books make up booked investment.
2. Annual, cumulative and average booked net income. The net profit (or loss) reported to stockholders is the booked net income.
3. Annual or average booked rate of return. The booked net income divided by the average net booked investment is the booked rate of return. It is often called the return on net assets. It can be shown for a single year, cumulative to some point in time, or an average over project life.

D. *Yardsticks related to risk*

So far we haven't mentioned risk. The next two chapters dwell on this subject. Risk may or may not be taken into account in calculating project economics. It is handled in different ways, depending upon the type of risk and the degree of risk.

1. Some risks are weighted for evaluation. We shall review risk weighting.
2. Other risks are not enumerated but are covered in an evaluation by a qualifying statement.
3. Risks are sometimes handled by yanking the minimum criteria high enough to offset (hopefully) the unknown risk.
4. The interrelation of numerous risks is evaluated by a mathematical technique often referred to as Monte Carlo simulation. It sums randomly the probabilities of each event occurring and produces a probability distribution for the combined risks. A computer is needed for this treatment of risks.

Investment yardsticks and the School prospect We have now discussed yardsticks and it is time to see them in application. We shall use the School prospect to review our knowledge of yardsticks.

Yardsticks related to cash flow First, let's review the total expenditures before and after tax. They were:

	M$	
	Total	*AFIT*
Investment	6,200	3,800
Exploration expenditures	3,000	1,900
Development expenditures	3,200	1,900

Note that the AFIT investment is shown as \$3,800,000. This number is used in cash flow analysis. The capitalized investment was \$3,600,000.

These expenditures are used with other data to construct additional yardsticks. The actual value profit (AVP) was \$5,200,000, the same as total booked net income. To get a profit-to-investment ratio for the School prospect, all we do is divide the net profit by the investment. The net profit used is always the AVP, but what investment do you use? The total investment? The capitalized investment? The after tax investment? Let's compare these for the School prospect.

Profit to investment ratios	*P/I*
Based on total investment	$\frac{5,200}{6,200} = 0.84$
Based on AFIT	$\frac{5,200}{3,800} = 1.37$
Based on capitalized investment	$\frac{5,200}{3,600} = 1.45$

Thirty years ago the net profit was most frequently related to the capitalized investment. It is seldom related to before-tax expenditures. Cash flow analysis has introduced a new term however—the ultimate net income ratio. Here the total net profit, ultimate net income, is related to the maximum negative cash flow, e.g., in the School prospect the ultimate net income ratio is:

$$\text{UNIR} = \frac{5,200}{2,100} = 2.48$$

Whenever the maximum negative cash flow is lower than the after-tax investment or capitalized investment, the ultimate net income ratio (UNIR) will be higher than the profit-to-investment ratio (P/I). In general this condition is true more often than not; therefore, the UNIR is usually higher than commonly used P/I ratios.

Yardsticks related to timing Next let's examine yardsticks related to timing for the School prospect.

Project life—years	13
Payout period—years	6.2
Discounted cash flow rate—%	22.2
From the PV profile—	
PV profit (M$) @ 8%	2,040
PV profit (M$) @ 10%	1,583
PV profit (M$) @ 20%	189

Yardsticks related to accounting procedures Although not used extensively in investment analysis, it is advisable to understand those yardsticks related to accounting procedures. For the School prospect these are:

Capitalized investment—M$	3,600
Booked rate of return—%	32.4
Total booked net income—M$	5,200

A comparison of investments With all these yardsticks how do you sort the "wheat from the chaff?" Remember we said there is no ideal yardstick; but what are the important ones? We can best answer this question by a comparison of three investments. Such a comparison is shown in the following table.

TABLE 25

Comparison of investment yardsticks

	Opportunity		
Yardstick	*A*	*B*	*C*
Investment, after FIT* $	300	300	300
Actual value profit $	**1,650**	1,140	300
Project life-years	35	15	**2**
Ultimate net income ratio	**5.5**	3.8	1.0
PV profit, discounted @ 10% $	300	**405**	222
Payout-years	5.0	3.3	**1.0**
DCF rate (IIR) %	21	35	**58**
Booked return %	**75**	60	50

* Maximum negative cash flow in this example.

In Table 25 we compare three investment opportunities, each with different income characteristics but with an identical initial investment of

$300. Investment A has the largest AVP of $1,650. By this standard it is the best. Investment C has the shortest project life. You get your money back quickly also with its 1.0 year payout. A, however, has the highest UNIR, but B has the largest PVP at 10%. C has the largest DCFR of 58%, but A has the biggest booked return of 75%. Thus, each investment has one or more yardsticks where it exceeds the others. Which is the best investment?

There is no clear cut answer to this question. We must know other related facts. For example, what are corporate objectives? If the corporation wants any investment to make a DCFR of 20% or more then all three qualify. If the company is not capital limited, it makes all three investments. Suppose it can make only one. Which one would you make? Again, it depends on another factor. If you could make C first and still take advantage of A or B one year later that would be the best route because of C's fast payout. If you can only make one and the other two won't be available in one year, then you must choose between A and B since they give greater profits for a single investment.

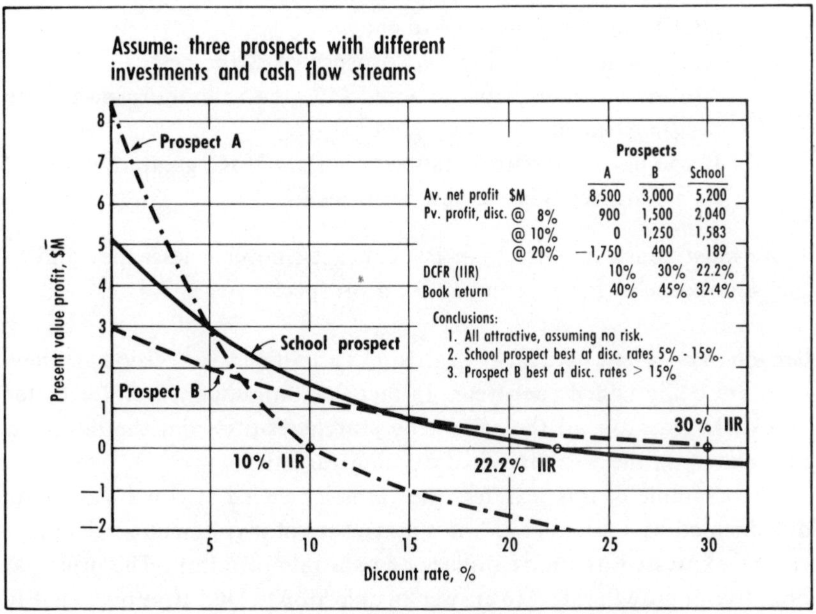

FIG. 11 COMPARING INVESTMENTS WITH PRESENT VALUE YARDSTICKS

If you don't have a lot of alternate investments and want the most AVP for the investment, take A; but B is better if you will have another 30% opportunity before the end of five years.

We haven't mentioned risk. Before a final decision could be made, you must consider the element of risk.

Comparisons using present value yardsticks One can use the shape of a PV profile to help understand a single investment and to compare two or more investments.

Fig. 11 is such a comparison. A few additional yardsticks are shown for each prospect on the right hand side.

Prospect A is obviously a long life project and it has the highest AVP of $8,500,000. Prospect B has a shorter life and a lower AVP but higher discounted profits than A at 8, 10, and 20%. Similar, now familiar, characteristics for the School prospect are also shown.

Here is what Fig. 11 tells us if we use only PV concepts as our investment criteria:

1. Assuming no risk and a minimum guideline no lower than 10% DCFR, all are good investments.
2. At discount rates below 5%, prospect A is the best.
3. At discount rates between 5 and 15%, the School prospect is the best investment.
4. If you have alternate investments with DCFR's greater than 15%, then prospect B is the best investment.

We must consider more than PV concepts; nevertheless, they add to our storehouse of information about a prospect.

Growth rate—a new yardstick Yardsticks are subject to revision and new ones are being added each year. In fact, a continuous search for better yardsticks goes on all the time. The search results from the desire to improve upon the weaknesses of existing yardsticks.

One example of this search is the yardstick called "Growth Rate." As documented by Capen, *et al*[2], it is a measure of investment efficiency. It can be expressed as future dollars or as a rate of return. The principal objective of growth rate is to answer an objection to DCFR. A high DCFR usually is well above a company's average return on investment. Therefore, cash flows received in future years cannot really be reinvested at the high rate indicated by the DCFR.

Growth rate attempts to answer this flaw by saying in effect "why not invest all future positive cash flows at the average company return?" This solution uses the "real world" rate of return and measures the growth of future cash flows to some future target year.

As a measure of investment efficiency, growth rate can be expressed as the present worth per dollar invested. However, this expression introduces yet another PW/I ratio to a host of such ratios and a percentage, if meaningful, can be grasped more quickly. For this reason a growth rate of return can be calculated, yielding a percentage as the final answer. If you are uncertain as to which yardstick answer would be the most easily understood, supply the requester with the PW/I ratio and the percentage.

Calculating growth rate of return How does one calculate this yardstick? The growth rate of return (GRR) is calculated by the following steps:

Step	*Action*
1	Compound (at the average company reinvestment rate) all *positive* cash flows forward to some target year (t).
2	Positive cash flows beyond the target year are discounted back to the target year.
3	Negative cash flows are discounted to the present at the same rate used in step one. The sum of the *discounted* negative cash flows is considered the investment (I).
4	The sum of the cash flows in steps one and two is designated as B.
5	Calculate the GRR as follows:

$$GRR = \left(\frac{B}{I}\right)^{\frac{1}{t}} - 1$$

where:

GRR = the growth rate of return (multiply the decimal answer by 100 to get a percent)

B = the sum of positive compounded cash flows to time t (target year) and the discounted cash flows beyond t

t = target year

I = the sum of the negative cash flows discounted to year zero.

Examples of GRR combinations The various combinations of factors influencing growth rate are demonstrated on Table 26. In column one is the net cash flow stream for the School prospect. It totals $5,200,000 but is composed of the negative stream of $2,100,000 and the positive stream of $7,300,000.

Columns 3, 4, and 5 show the cash flows with three different target years. In each column the cash flow has been compounded or discounted appropriately as outlined in the first four steps of the GRR calculation. For simplicity, annual compounding was assumed and the annual cash flow was assumed to have been received at year end. The reinvestment rate was assumed as 10%.

In column three, the target year (t) was 8 with the calculation of GRR as follows:

$$GRR_{10} = \left(\frac{7179}{1766}\right)^{\frac{1}{8}} - 1 = .192 = 19.2\%$$

Column 4, with t = 10, calculates to a GRR of 17.3% and column 5 at t = 13 has a GRR of 15.5%. Varying the target year makes a difference. However, according to Capen, *et al* the relative relationship between investments will not change with change in t; therefore, it does not lose its discriminatory value.

In columns 3, 4, and 5, the discounted value of negative cash flows was used as the investment. Just for comparison, column 2 shows the positive cash flows compounded to t = 8 but with the negative cash flows not discounted.

In this case:

$$GRR_{10} = \left(\frac{7179}{2100}\right)^{\frac{1}{8}} - 1 = .167 = 16.7\%$$

In other yardsticks in this chapter, "I" was never less than the maximum negative cash flow. The GRR, using the maximum negative cash flow, was calculated merely for comparison purposes and is *not recommended.* According to all cash flow theory and the theory of discounting, all investments (in this case the negative cash flows) should be discounted to

TABLE 26

Various calculations of growth rate
School prospect
M$

	1	2	3	4	5
	Net	Growth rate @ 10%*			
	cash	I=2100	I=1766		
Year	flow	t=8	t=8	t=10	t=13
0	−500	−500	−500	−500	−500
1	−200	-200	−182	−182	−182
2	−600	−600	−496	−496	−496
3	−600	-600	−451	−451	−451
4	−200	−200	−137	−137	−137
5	700	932	932	1,127	1,501
6	1,200	1,452	1,452	1,757	2,338
7	1,200	1,320	1,320	1,597	2,126
8	1,200	1,200	1,200	1,452	1,933
9	600	545	545	660	878
10	600	496	496	600	799
11	600	451	451	545	726
12	600	410	410	496	660
13	600	373	373	451	600
Totals	+7,300	+7,179	+7,179	+8,685	+11,561
	−2,100	−2,100	−1,766	−1,766	−1,766

*Assume income received on last day of the year shown.

year zero. We assume that future investment money is at work earning interest until needed.

What have we learned from these calculations? It is obvious that GRR increases, at a constant I, as t becomes smaller. Capen, *et al* demonstrated that GRR does increase as t approaches 0. As t increases, GRR approaches the reinvestment rate.

For the School prospect this relationship is confirmed on Figure 12. Assuming we wish to have a target year beyond the last negative cash flow, then t can be as small as 5. The thirteenth year is the last cash flow year and can also be a target year. The plot on Figure 12 demonstrates the range of GRR from t=5 to t=13. The GRR, expressed as a percent varies from 25.0 at t=5 to 15.5% at t=13. So it is possible to have GRR values both above and below the DCFR value of 22.2%. At a target year of 6 the GRR is 22.4%, almost the same as the DCFR.

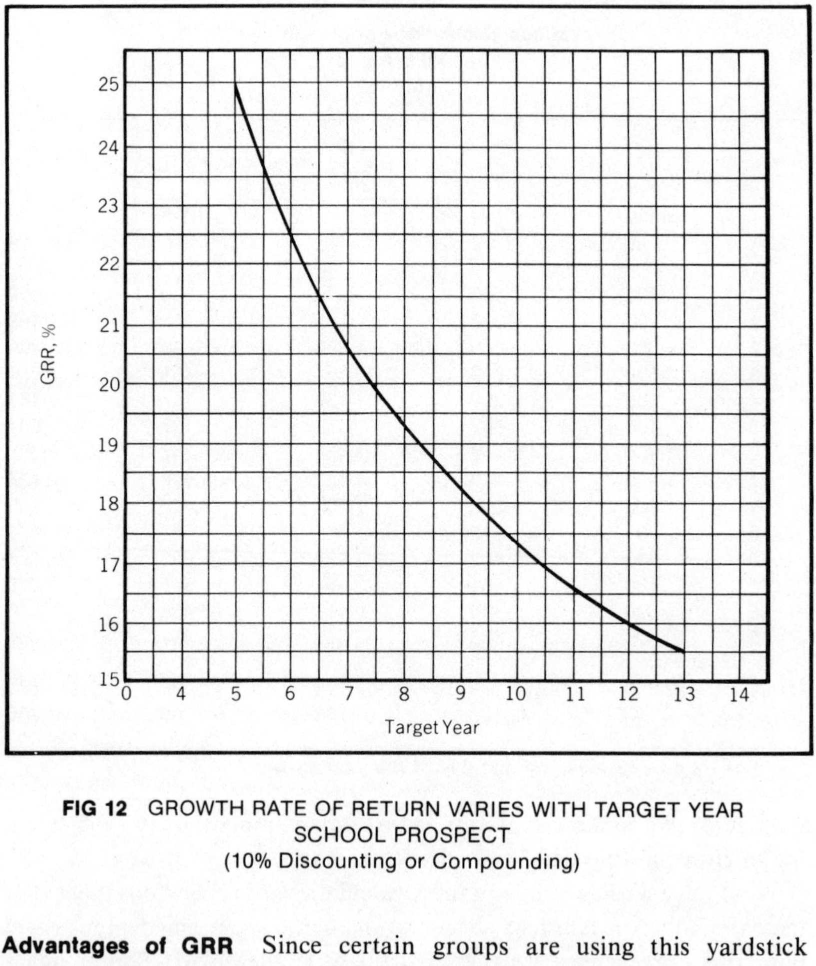

FIG 12 GROWTH RATE OF RETURN VARIES WITH TARGET YEAR
SCHOOL PROSPECT
(10% Discounting or Compounding)

Advantages of GRR Since certain groups are using this yardstick preferentially, they must consider it to have advantages. What are these advantages?

1. Growth rate is not a trial-and-error process as is DCFR. Therefore the calculations are easier.
2. Growth rate shows what the company can expect if the future cash flows are reinvested at its average rate of return.
3. Growth rate of return is an extension of present worth per dollar invested. It produces a percentage for those accustomed to a percentage to compare investments.

Are there disadvantages? Although they might not qualify as severe disadvantages, several key decisions must be made relative to GRR.

1. The choice of the time horizon (t) must be made. The final choice won't change project ranking by GRR, so what target year should you use? Probably you should use one which produces percentages close to those normally seen by your management.
2. The reinvestment rate must be chosen. The problems here stem from varying functional rates of return. If your function consistently outperforms the other functions of the company, the average company reinvestment rate may be below what you should be using. Some companies have functions for which no reinvestment rate is calculated. In these cases, a borrowing rate might be used. Whatever the difficulties, a decision must be made on the reinvestment rate.
3. The definition of investment must be considered. There should be few problems with the recommendation of Capen *et al* to use the discounted negative cash flows as I. Nevertheless, this I is less than almost all other concepts of investment.

The extended discussion of growth rate and GRR illustrates the search for more definitive yardsticks. For exploration investments one should make sure that risk and risk alone does not "swamp" all minor differences in yardstick percentages. Our risks are still primarily geologic ones. This observation should in no way detract from the search for more meaningful yardsticks.

Other evaluations We began this section by stating that yardsticks can be revised. Several new innovations result from the modifications of existing, long-used yardsticks.

An interesting modification of payout was proposed by Robert A. Campbell and J. M. Campbell, Jr.[3] It is *discounted* payout. The logic behind the change is to require the investment to return not only the principal but also the interest required to make the investment.

Campbell and Campbell also list other new yardsticks which are being tested. Most incorporate the discounted payout (DPO) as part of the yardstick. A brief description of these yardsticks follows:

Treasury worth (*TW*)—TW is computed as the discounted cash flow before DPO plus the undiscounted cash flow after DPO.

Treasury worth index (TWI)—An index of the discounted cash flow before DPO plus undiscounted cash flow after payout (TW) divided by the investment.

Accrued treasury assets (ATA)—ATA is the sum of discounted cash flow until DPO, plus the accrued value of all cash flows after DPO. ATA differs from TW in that it develops an equivalent future value after payout. In contrast, TW develops an equivalent present value.

Treasury growth rate—This yardstick utilizes the ATA concept to develop a rate of return. To calculate TGR you compound all cash flows after DPO, then discount all cash flows prior to DPO. Campbell and Campbell show the relationship between ATA and TGR as follows:

$$I\ (1+\mathrm{TGR})^n = \mathrm{ATA}$$

where n is the year of the last cash flow—the end of the project.

In summary, we see that TW is related to TWI in much the same manner as ATA is related to TGR. For comparing investments TGR can be given a common reference point (n) for a more equitable accept/reject criterion.

Yardsticks are changing and will change as new conditions emerge. We should expect change and be willing to test new concepts in measuring one investment relative to another.

REVIEW

You have related factors, beyond yardsticks alone, that must be considered to choose the right investment. They include:

1. Risk
2. Corporate objectives
3. Number of available opportunities
4. Available capital

Despite these limitations you can see that the various yardsticks tell you much of what you need to know about an investment opportunity. They do not make the decision, but they greatly assist the decision maker.

The explorationist probably has more independent factors to consider than any other investment maker. His experience and judgment are vital to proper choices. Investment analysis can be one of his greatest tools and it will show him the economic consequences of his geologic assumptions.

BIBLIOGRAPHY

1. Wooddy, L. D., and Capshaw, T. D., "Investment Evaluation by Present Value Profile." Paper presented to the SPE annual meeting, October 1959.
2. Capen, E. C., Clapp, R. V., and Phelps, W. W., "Growth Rate—A Rate of Return Measure of Investment Efficiency." Journal of Petroleum Technology, May 1976, pg. 531.
3. Campbell, Robert A., and Campbell, J. M., Jr., "Optimizing Treasury Growth With Proper Evaluation of Long Term Projects." Paper presented at SPE Meeting, February 1977.

8 *Risk Analysis— Fundamentals*

You might think at this point that you are ready to embark on the evaluation of exploration investments; but we haven't discussed one of the most important items in our evaluations—risk. It is the aim of the next two chapters to take economics out of cost, where we've been for almost seven chapters, into risk.

The most important single variable in the evaluation of exploratory investments is risk. It is the one factor about which others can tell you the least. In the search for hydrocarbons, facts are few and decisions must always be made before all facts are known. Because of the many unknown factors, we depend a lot on experience in exploration. But the wise explorer knows that the facts, though few, are always worth reviewing.

Only two chapters in this book are devoted to risk analysis. In reality the subject of risk analysis deserves much more space. However, any introductory text must limit discussions to basic fundamentals. For those interested in an expansion of Chapters 8 and 9, a companion volume is available. It is entitled "An Introduction to Risk Analysis" and was published in 1977.[1]

Risk vs. uncertainty Many people equate risk to uncertainty. In this chapter we will make a distinction between the two. We shall always use risk to mean an opportunity for loss. The term uncertainty shall apply to factors where the outcome is not certain but where the opportunity for loss is not as apparent as in risk.

The chances for loss are easy to distinguish in an exploratory well. Statistics alone show loss to be more normal than gain. In an investment evaluation, however, there are numerous uncertainties regarding such things as legislation, interfuel competition, etc. These factors affect the economics but few have the outright opportunity for loss possible in exploratory risk.

Quantifying risk in exploration If you put down a rough formula to describe the economics of exploration, it might look as follows:

$$\text{Hydrocarbon potential} + \text{economics} + \text{risk} = \text{decision}$$

You may wish to substitute prospect or play for the first item; but regardless of the first term, we take geological-geophysical concepts involving hydrocarbons, determine the economic consequences of these concepts and apply risk. To this point we have assumed hydrocarbon potential, shown methods for calculating economics, and we now approach the subject of risk.

We shall show ways to quantify the risks of exploration. In so doing, we always face the possibility of quantifying ignorance, or bias, or both. However, decisions are made every day in exploration based on mental quantification. Such quantification relates to one man's experience. His "long shot" may be another's "fair chance" or another man's "risky." We need some standard method to gain comparable definitions of risk for the same prospect. Quantification at least offers the hope of more consistency in expressing risk. It should also encourage the quantifier to review his basis for establishing risk and in so doing reduce errors of oversight.

Risk defined by probabilities Before delving into exploratory risks, two fundamental laws of probability will be briefly reviewed.

Probability theory considers all possible happenings and attempts then to estimate the most likely happening. Even though the theory usually has meaning only if there are a significant number of happenings, it can teach us important things about few happenings. It can help us estimate the odds.

As we have found in other chapters on other subjects, some new terms are required. In probability theory, we need to know the following:

An *event*—is one of two or more things which may occur in a given situation.

An *outcome*—is the thing which does occur. One thinks in terms of events before the fact and outcomes after the fact.

Probability—is a statement of the frequency of a specific event being the outcome after repeated trials. It is expressed as a per cent or a fraction.

The fundamental rules of probability are:

The addition rule and the multiplication rule. Both rules will be illustrated by rolling dice.

The addition rule What is the probability of rolling a five on one roll of one die?

There are six possible events: 1, 2, 3, 4, 5, and 6. So the probability of rolling a five is one in six or one-sixth ($\frac{1}{6}$). Expressed as a percentage there is a 16.7% chance of rolling a five in one roll of one die. Note that when any one of the six possible events does occur it excludes the other five possibilities. These events are thus mutually exclusive. The addition rule says that the probability that one or another of two or more mutually exclusive events will occur is the sum of their separate probabilities.

We can review the rule by considering more than one event. What is the probability of rolling a two or a five on one roll of one die? Reading P(5) as the "probability of five" our question is:

$$P(2 \text{ or } 5) = ?$$

The two events are mutually exclusive, so we can apply the addition rule.

$$\begin{aligned} P(2) &= \tfrac{1}{6} \\ P(5) &= \tfrac{1}{6} \\ P(2 \text{ or } 5) &= P(2) + P(5) \\ &= \tfrac{1}{6} + \tfrac{1}{6} \\ &= \tfrac{1}{3} \\ &= 33.3\% \end{aligned}$$

The multiplication rule When two dice are rolled, the outcome of either die is not affected by the other. The outcomes are *independent* of each other. The multiplication rule states that: the probability of two independent events having specific outcomes is the product of their separate probabilities.

Consider the question, "What is the probability of rolling a double five on one roll of two dice?"

$$P(5,5) = ?$$

The event of a five showing on one die is independent of a five on the other die. Thus the multiplication rule applies.

$$\begin{aligned} P(5) &= 1/6 \\ P(5,5) &= P(5) \times P(5) \\ &= 1/6 \times 1/6 \\ &= 1/36 \\ &= 2.8\% \end{aligned}$$

Probabilities and the search for hydrocarbons The search for hydrocarbons is by its very nature a function of complex probabilities. One can summarize these probabilities in a threefold generalization.

Geologists and geophysicists are concerned first of all with the probability that hydrocarbons do exist in a specific geographic area. Second, they are concerned with the probability of being able to find these hydrocarbon deposits. Third, they must be concerned with the probability of whether the deposits found will be economic.[2]

Are hydrocarbons present? In every unexplored basin, geologists have been faced with the basic question, "Are hydrocarbons present?" The only final answer to this question is the wildcat well; but in order to justify an exploratory well, the geologist looks for reasons why hydrocarbons could be present. He may try comparing the unexplored basin to a similar but explored basin. He will look for optimum sedimentary depositional environments, lithologic seals, trap forming mechanisms, etc.

We will deal very lightly with economic approaches to this basic question (see Chapter 10). The assessment of an unexplored basin must begin as a geologic interpretation and economic calculations can be handled simply until more data are available.

Will you find them? After oil or gas is discovered, the question becomes, "If hydrocarbons are present, will you find them?" The answer to this question involves much more than technical skill. Land ownership, degree of competition, available capital, timing of lease sales and many other factors affect what a given concern can accomplish. Technical skill without the backing of land and money to explore the land is of no value.

If found, are they economic? The probabilities here involve size, amount of revenue generation possible (deliverability) and costs. Chapter 10 will deal extensively with combinations of these factors. Later in this chapter we will comment on field size. For now we shall return to what we can call "wildcat" risk. Chapter 9 deals with "economic risk."

Wildcat risk If we modify our general formula to: Prospect + economics + risk = decision, several significant illustrations involving risk and probability are possible. We chose the prospect because most of the explorationists' efforts relate to specific prospects. Our definition of wildcat risk, then, is the chance a prospect has of making a discovery.

For illustration purposes we shall use a very simplified method for quantifying wildcat risk. A few very complex procedures have been suggested. A recommendation by Schwade[3] includes five pages of a prospect check list. His suggestions pertain to the geologic factors which make up wildcat risk. No one knows exactly what conditions produce the most favorable chance of all geologic factors acting favorably; because of this fact, we will never have a uniform agreement on what constitutes wildcat risk.

There are numerous obvious factors affecting the trapping of hydrocarbons and one's ability to map them. Some important examples are: subsurface control, type of trap, quality of basic data, nearness to production or shows, and many others. An ideal system would isolate all pertinent factors, assign probabilities to them and combine them into a composite risk factor. Our simplified method can be made more complex. It is used herein for illustration purposes only. Anyone interested in additional research on prospect risking will find ample reference in the literature.[4,5]

Simplified method for estimating composite geologic risk Our method will concern itself with three factors only, which shall be assumed to encompass all geologic conditions favorable for hydrocarbon entrapment. These factors are:

1. *Structure.* This factor shall encompass faulting, type of trap, seal, geophysical control and quality, subsurface control, etc.

2. *Reservoir.* This factor shall include rock properties, depositional conditions, and other factors specifically affecting a productive reservoir.
3. *Environment.* This factor shall include paleohistory, timing aspects, proximity of source beds, shows, etc.

One final thought on factors affecting wildcat risk. Data will vary from prospect to prospect and an absolutely uniform system cannot always be used. In the absence of adequate data you may settle for estimating only the first two—structure and reservoirs. If you cannot reasonably assess a factor, leave it out. Note the modifying term "reasonably." You are always dealing in uncertainties in estimating probabilities; but you must assess some factors (say structure and reservoirs) otherwise you may have no prospect.

We are now ready to illustrate the developing of a number which will represent a geologist's best estimate of a prospect's chances of being a discovery. We shall consider our three factors more or less independent of each other. Remembering the multiplication rule, our steps to estimate wildcat risk become:

1. Isolate critical geologic factors (which we have defined here as three).
2. Estimate each factor's chance of being favorable.
3. Multiply the per cent chances of all factors together.

The product is an estimate of the overall chance of success; it is an estimate of the probability of all geologic factors occurring favorably simultaneously.

What can you learn from past wildcats? If you have never considered this approach, you may wish to conduct some useful research first. Check the last 10 to 20 rank wildcats drilled. Fifty would be a better sample, but you may not have drilled that many. We shall assume you have made such an analysis and use the results to demonstrate our method of risking. Suppose your research of 50 wildcats results in the following data:

Wells finding	*No. of wells*
Structure as predicted	35
Good reservoir conditions	25
Optimum environment	41

These figures tell us that:

1. Structural predictions were correct 70% of the time.
2. Only 50% of the time were good reservoirs present.
3. About 80% of the time the proper environment was present.

Your historic wildcat risk should be about (remember the multiplication rule):

$$0.7 \times 0.5 \times 0.8 = 0.28 = 28\%$$

If this factor is near your success rate (if you have 14 discoveries) your data reasonably measured the critical geologic factors.

Your research makes two important contributions. If data were sufficient you now have a glimpse of history and its relevant probabilities. If your data were insufficient you need to begin recording estimates and critical factors. It is by carefully reviewing what was estimated under uncertainty that you can hope to sharpen your predictions.

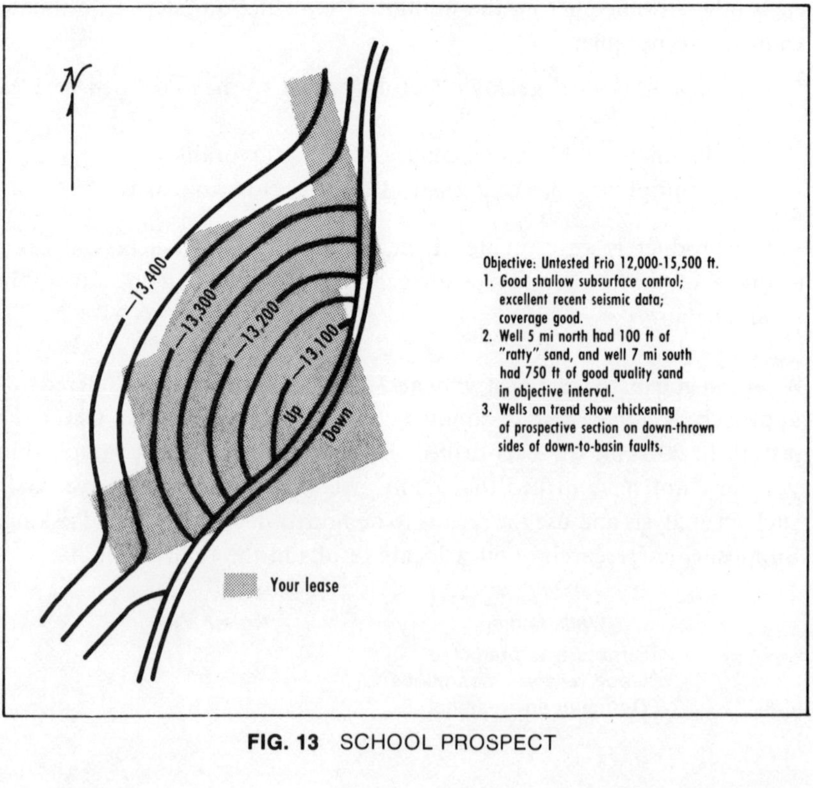
N
-13,400
-13,300
-13,200
-13,100
Up
Down
Objective: Untested Frio 12,000-15,500 ft.
1. Good shallow subsurface control; excellent recent seismic data; coverage good.
2. Well 5 mi north had 100 ft of "ratty" sand, and well 7 mi south had 750 ft of good quality sand in objective interval.
3. Wells on trend show thickening of prospective section on down-thrown sides of down-to-basin faults.
Your lease

FIG. 13 SCHOOL PROSPECT

Method applied to a single prospect One can apply the technique to a single prospect to estimate its probability of having favorable geologic conditions. We shall again use the School prospect shown in Fig. 13 contoured on top of the objective formation.

Note the factual information on the right of the figure. Using these facts about the School prospect, an explorationist's thinking might go as follows:

> "Hum... the structure is tied down well so it should have a 75-80% chance of being as predicted. Now the reservoir conditions look like a fair shot; however, at this depth our experience in finding good sand has been poor—maybe a 40-50% chance? The Frio is a known producing formation in the area, although this particular fault segment has turned up nothing large so far. Timing and migration, no problems here—say 85% chance."

The foregoing thoughts could be translated and recorded on an appraisal form as follows:

Wildcat appraisal

Prospect	School	Proposed T D	15,500 feet
Location	Nonesuch	Objective	Frio
		Interval	12,000 feet—TD

Risk estimation

1. Structure	50, 60, <u>70</u>, 80, 90	77	Composite
2. Reservoirs	30, 40, <u>50</u>, 60, 70	45	risk
3. Environment	60, 70, <u>80</u>, 90, 100	85	factor
			0.29*

*0.77×0.45×0.85=0.29

The composite risk factor is now a synthesis of the geologist's judgments about critical factors. He has had to review his data and record his opinions. Note that the form has numbers in the middle column with one number underlined. The underlined numbers are those from your research and are placed on the form for relevance but not as absolute guidelines. One's ability to predict critical factors may change, you may be estimating in a different trend, in a relatively unexplored area, etc. However if your estimates consistently depart from prior historic results, you need to examine the cause. You are either establishing new history or your estimating may not be in the ballpark.

The composite risk factor is expressed as a single figure—0.29. There is probably no risk difference between 0.27 or 0.29; furthermore it is difficult to say how much difference is required to be meaningful. When you have numerous prospects to appraise, however, a method of establishing a composite risk is useful and necessary for risk ranking. The estimate is subjective and we shall find uses for its quantification in our calculations involving economic risk to be shown in Chapter 9.

Before proceeding to economic risk, two additional facets must be examined. They are statistical success and field size distributions. Both terms relate to techniques which have applications in making decisions on exploration investments.

Statistical success Explorationists usually ponder how often success will be achieved from a given program of wildcats. A company drilling 20 or 30 wildcats per year might want to know the odds of making one, two, three or five discoveries. Keep in mind that the word discovery used here means simply a producing well. How much is discovered will be discussed under field size distributions.

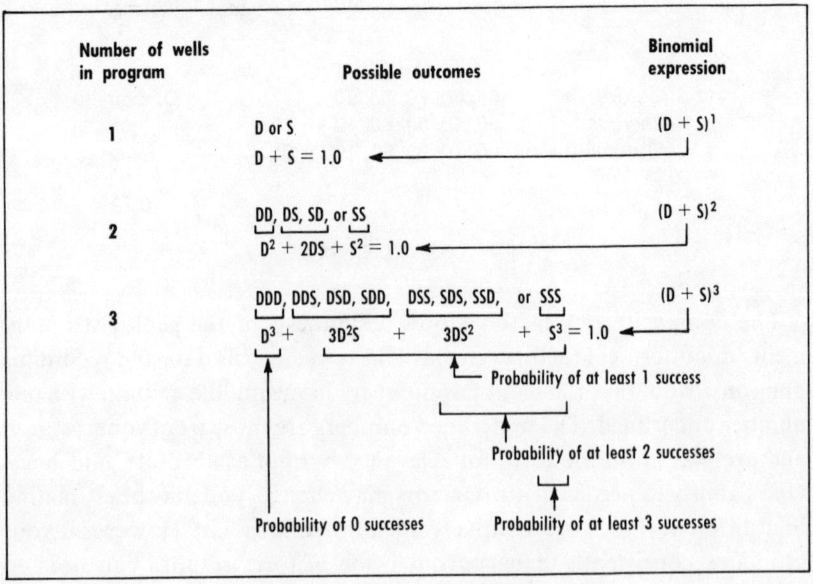

FIG. 14 BINOMIAL EXPANSION

To find these odds a mathematical technique called binomial (two numbers) expansion is used. To simplify matters we shall assume each well in the program has the same chance of success. As explorationists we know that some prospects have better odds than others; but for most exploration programs, we can assume an average success rate with reasonable safety.

Our symbols will be:

D = Probability of a dry hole
S = Probability of success

For one well

$$D + S = 1.0 \qquad \text{Equation 9}$$

We can also write

$$D + S = (D + S)^1$$

For a two-well program, see Figure 14; there are four possible outcomes and our equation becomes:

$$DD + DS + SD + SS = 1.0 \qquad \text{Equation 10}$$

Of course DS + SD can be written as 2 DS which allows us to write the probabilities for two wells as follows:

$$D^2 + 2DS + S^2 = 1.0 \qquad \text{Equation 11}$$

If you remember your algebra, $D^2 + 2\,DS + S^2$ is the product of $(D + S) \times (D + S)$ and can be written as $(D + S)^2$. Therefore,

$$D^2 + 2DS + S^2 = (D + S)^2 \qquad \text{Equation 12}$$

The left half of Equation 12 is the expansion of the binomial (D + S) to $(D + S)^2$.

The mathematics associated with binomial expansion is easy only if the numbers are small or simple. For example if you assume D and S have an equal chance (50% probability) and you drill only two wells, then DD + DS + SD + SS tells us that:

You have a 25% chance of drilling two dry holes (DD).
You have a 50% chance of drilling only one successful well (DS + SD).
You have a 25% chance of drilling two discoveries (SS), and
You have a 75% chance of at least one discovery (DS + SD + SS).

However, when you go to larger numbers of wells and other probabilities for D & S (success rates), the math becomes much more complex. Consider the binomial expansion for a 5-well program which is:

$$D^5 + 5\ D^4\ S + 10\ D^3\ S^2 + 10\ D^2\ S^3 + 5\ DS^4 + S^5 = 1.0$$

Fortunately for all of us most textbooks on statistics or manuals of statistical series have tables on binomial probabilities related to the number of trials (number of wells in a program). The tables show individual probabilities (the chance of exactly x successes in an N well program) or cumulative terms (the chance of x or more successes in an N well program). From these tables graphs can be constructed to show the odds for various numbers of discoveries, given certain wildcat success rates. Figure 15 is such a graph. It assumes a 10% success rate. If we have a 20-well program, the graph tells us that our chance of:

At least one (one or more) discovery	is 88%	88 in 100
At least two discoveries	is 61%	3 in 5
At least three discoveries	is 32%	1 in 3
At least four discoveries	is 13%	1 in 8

The chance of drilling any number of dry holes *in succession* can be determined from the "≥ 1 discovery" curve (≥ means equal to or greater than; or you can think of it as the curve showing the chances of *at least* x discoveries—in the case referred to here—the chance of at least *one* discovery). The chance of one dry hole "in succession" is (1.00 − 0.10) or 0.90 or 90%. For additional wells read the value on the ≥ 1 curve and subtract from 1.00; examples are:

2 dry holes in succession = 81%	4 in 5
5 dry holes in succession = 59%	3 in 5
10 dry holes in succession = 35%	1 in 3
20 dry holes in succession = 12%	1 in 8

One can state the last point another way. At a 10% success rate, even in drilling 20 holes we still face a 12% chance that all are dry.

These illustrations should give you some idea of the limitations of a single success percentage, even from only a mathematical viewpoint. The binomial expansion also provides additional information about risks at given success rates. It is from binomial expansion that we get concepts about required exposure for a reasonable probability of success and insights into "Gambler's Ruin" concepts.[1]

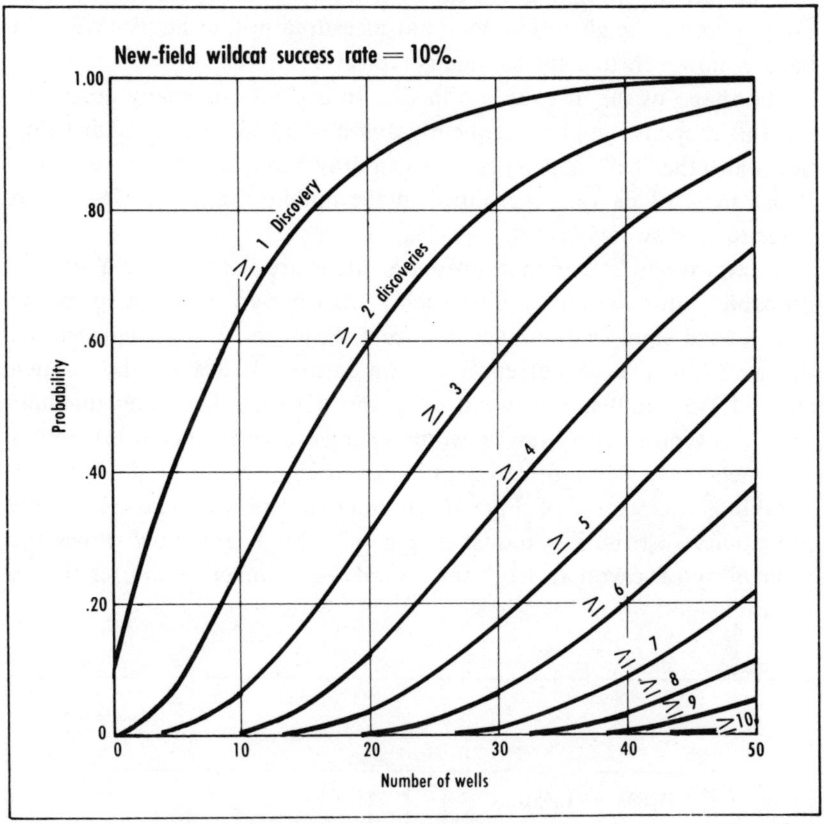

FIG. 15 CUMULATIVE BINOMIAL PROBABILITY

The normal distribution Now some comments on distributions. A distribution is an orderly arrangement of similar things.

The arrangement relates to a common characteristic of interest. Fig. 16 is frequency distribution of the heights of a sample of women. In this plot the group of similar things is women and the common characteristic is height. The bar graph of Fig. 16 is sometimes called a histogram (history + graph). Histograms plot class intervals versus their relative frequencies. In Fig. 16 the class interval is the height group and each bar shows the number of women in that group. For example, the frequency of occurrence of women whose heights range from 61½ to 62½ (in this sample of 1,345 women) is 186.

If we feel this sample is typical of all other women, the chart can be used to predict heights of women not measured in the sample. We must believe, however, that the sample is typical.

The shape of the distribution in Fig. 16 is similar to many others. In fact this shape is found so frequently that it is called the normal distribution—also the "bell shaped" curve when drawn as a continuous function. It has most of its values grouped in the middle range with fewer and fewer toward two extremes.

If percentages rather than absolute values are used on the Y axis, a probability distribution results; such a distribution is indicated by the scale to the right in Fig. 16. This scale shows the Y axis representing the probability of occurrence. In the sample 13.8% of the women ($186 \div 1{,}345 \times 100$) are between $61\frac{1}{2}$ and $62\frac{1}{2}$ in. Assuming that our sample is typical, every time a woman is measured for height, there is a 13.8% chance it will fall between $61\frac{1}{2}$ and $62\frac{1}{2}$ inches.

Adding the values of Fig. 16 cumulatively produces a cumulative probability distribution shown in Fig. 17. The Y axis now shows the probability of a woman from this sample being taller or shorter than a certain height.

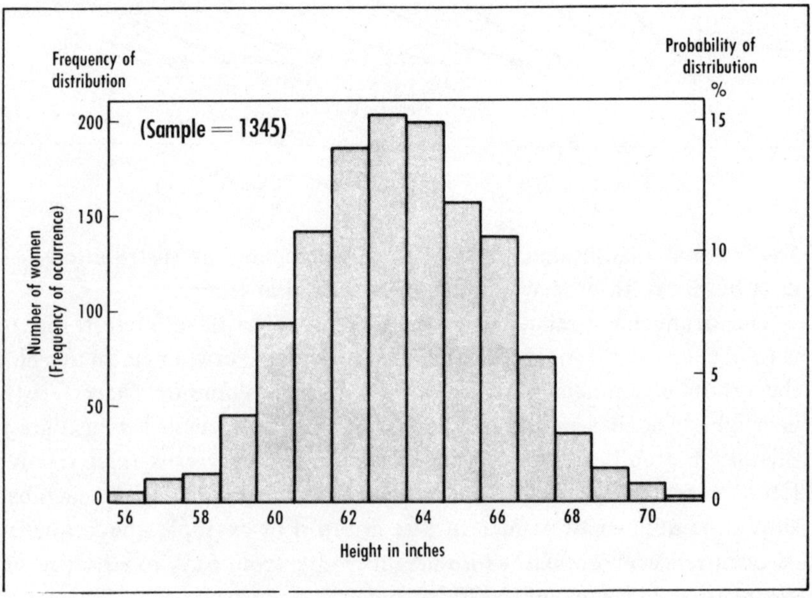

FIG. 16 HEIGHTS OF WOMEN

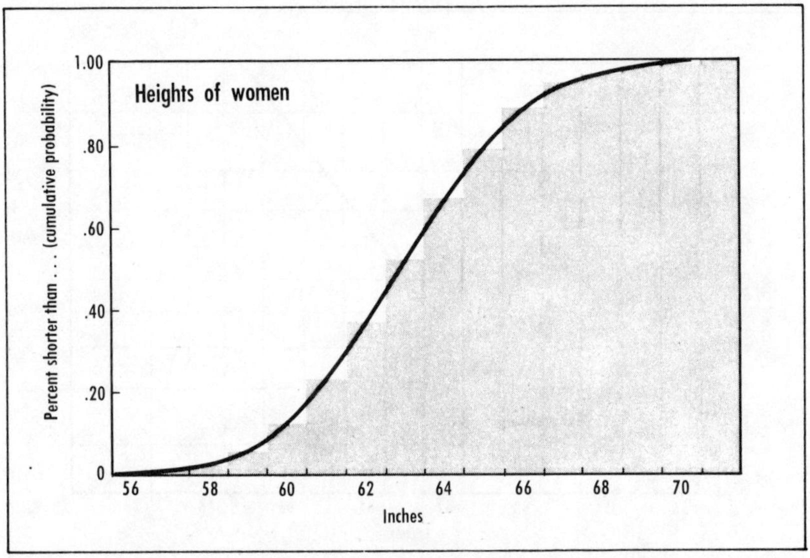

FIG. 17 CUMULATIVE PROBABILITY DISTRIBUTION

Straight lines are easier to work with than curved lines and a particular kind of graph paper is available for probability plots. A probability grid is shown in Fig. 18. The Y axis has been changed in such a way that cumulative probability values of normal distributions fall in a straight line.

All these comments have been leading toward a particular distribution useful to explorationists.

Nature's group of fields One of the basic reference points in a mature trend is the size of existing fields.

It is well documented that, within a given geologic unit (basin or formation) field size exhibits a specific frequency distribution.[6] If you plot frequency of occurrence versus the logarithm of field size, the resultant histogram is similar in shape to a normal distribution. Since the logarithm of the variable is plotted, the distribution is called a log-normal distribution. This natural arrangement can be used to describe and understand the distribution of oil and gas accumulations within the earth's crust.

Constructing a field size distribution (FSD) plot First, however, let's demonstrate, in a very simple way, how one constructs a FSD plot. The steps are as follows:

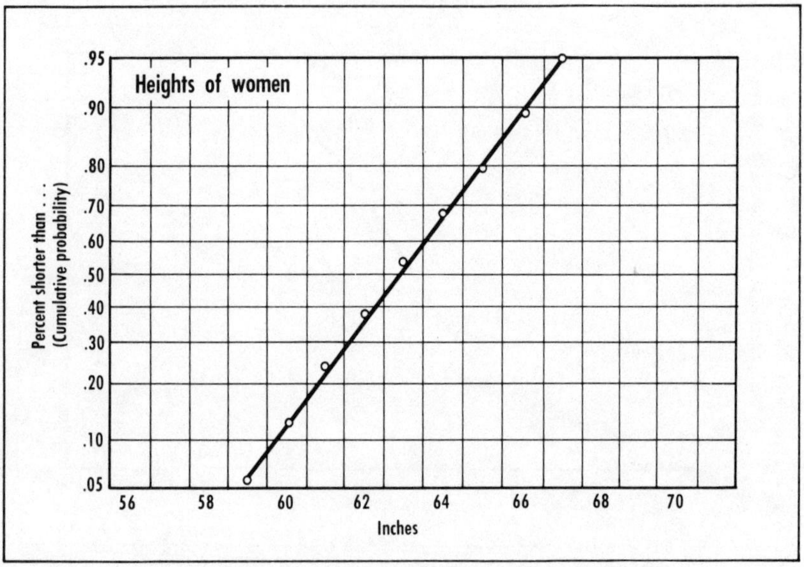

FIG. 18 CUMULATIVE PROBABILIY DISTRIBUTION

1. Arrange all fields by size with the largest listed first, the smallest last.
2. Calculate the per cent each field is of total fields.
3. Add the percentages for each field to construct a cumulative percentage for fields—begin with the largest field.
4. Plot actual field size versus midpoints of cumulative percentage.

You should plot the midpoints when you have only a very small sample of the total population, because none of your fields can, by definition, be the largest or the smallest (i.e., you can have neither a zero nor a 100% plotting point). Kaufman recommends using fractiles rather than midpoints.[6] Fractile estimates are calculated by using the total sample + 1 as the divisor for cumulative percentages. When the sample has 25 points or more, fractile estimates become almost the same as actual cumulative percentages.

We can demonstrate this procedure with just a few fields. Given 5 fields of sizes 40, 160, 70, 10, and 22 BCF:

	Calculate				
		cumulative percentage (2 & 3)			
	(1) Arrange by size—BCF	*Actual*	*Midpoints*	*Fractile estimates*	*(4) Plot size vs. midpoints on log-normal probability paper*
1	160	20	10	16	
2	70	40	30	33	
3	40	60	50	50	
4	22	80	70	67	
5	10	100	90	83	

The above data are plotted on Fig. 19 which utilizes log-normal probability graph paper. A straight line on such paper indicates a log-normal distribution. Frequently such graphs are viewed with the paper in the vertical position. The data are the same, so if this position seems more normal, just rotate the paper 90°.

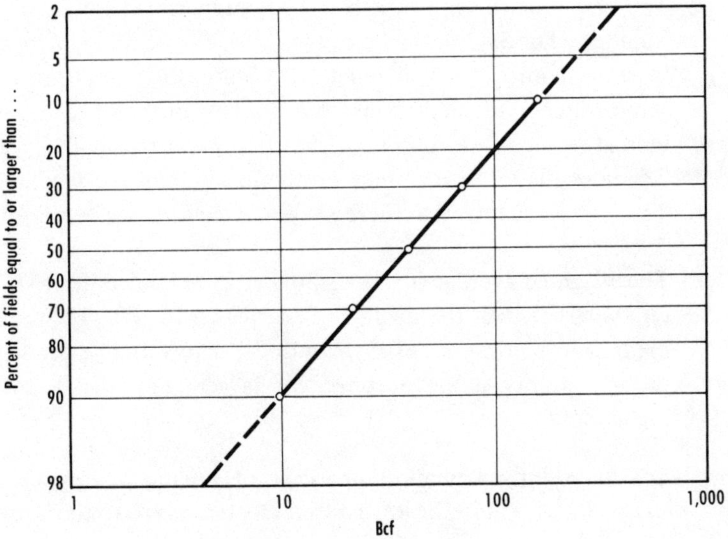

FIG. 19 FIELD SIZE DISTRIBUTION

Some facts about log-normal graph paper Other than the fact it shows a straight line, why use log-normal paper?

First, we know that nature most normally groups fields by this distribution if the sample is large enough. Thus we should plot field size data on log-normal paper to see if the sample is adequate.

Second, since we have a firm idea of the configuration of distribution, we can use this relationship to our advantage for implied or inferred distributions. When we begin to estimate prospect size in Chapter 9, some inferred distributions will be utilized.

Now, what can we observe about this distribution assuming that it truly reflects a larger sample and the entire population?

1. The median size is 40 BCF. One-half of the discoveries are larger than this size and one-half are smaller.
2. The construction of the graph paper is such that it clusters discoveries around the median. This clustering serves to demonstrate the size concentration. Note that 60% of the fields (between the 20 and 80% lines) will have sizes between 15 and 100 billion cu. ft.
3. The effect of the log plot can be demonstrated by these same numbers. The 30% of the fields from 80% to 50% shows a field size increase from 15 to 40 or a net of 25 billion cu. ft. In contrast the same number of fields from 50% to 20% probability shows an increase from 40 to 100 billion cu. ft. or an increase of 60, almost 2.5 times the volume change on the low side of the median.
4. The log plot means that there are fewer big discoveries than small ones.
5. The weighted average (mean) is almost 55 billion cu. ft., 15 billion cu. ft. larger than the median. The mean will always exceed the median in a log-normal distribution because the upper half of plotted points is proportionately larger, i.e., the X axis is logarithmic.

Using FSD plots for calculating chances of finding a specific field size Field size varies more or less continually from zero to the largest in the basin.

Because of this fact the chance of finding a field of a specific size is near zero. In spite of this fact, one can use FSD charts to determine the approximate chance of finding fields *near* specific values.

Remember of course that a FSD graph deals with discoveries only. Data from the graph must be multiplied times the success ratio to determine the chance of success.

Using the curve of Figure 19, what is the chance of finding a 50 billion cu. ft. or larger discovery in a play where the "going" success rate is 20%? From Fig. 19 we see that 40% of the fields are equal to or greater than 50 billion cu. ft. Then what are our probable chances—if we get a discovery? Since there's only a 20% chance you will be successful the chance of finding a 50 billion cu. ft. field or larger is:

$$.40 \times .20 = .08$$

Based on our FSD graph, only 8% of the time will a well find at least 50 billion cu. ft., or 2 chances in 25.

The chances of finding at least a 10 billion cu. ft. discovery (at 20% success) are:

$$.90 \times .20 = 0.18$$

You have slightly better than one chance in six of getting a discovery equal to or greater than 10 billion cu. ft. at a 20% success rate.

You may have only a small exploration program in an area or a trend. Furthermore, you may have your prospects already selected with no more acreage available. In such cases you are dealing with specific data, not samples of a large population. Field size distributions may, under these special circumstances, seem of little value. Nevertheless, FSD's are very useful in maintaining a touch with reality when estimating size; and if you plan exploration programs in mature trends, your view of your poten-tial discoveries will be more meaningful if you understand nature's distribution of oil and gas fields in that play.

REVIEW

This chapter covers a few basic concepts associated with wildcat risk. Probability theory helps us combine geologic fact and opinion in ways which clarify our best estimate of risk. Where we have sufficient data, statistical analysis and field size distributions help us understand the odds.

We are now ready to combine risking techniques with economics to determine profitability. The combination gives us risk-weighted profitability.

BIBLIOGRAPHY

1. Megill, R. E., "An Introduction to Risk Analysis," The Petroleum Publishing Co., Tulsa, Okla., 1977.
2. Megill, R. E., "How to Measure Exploration Profits," The Oil & Gas Journal, March 18, 1968, p. 126.
3. Schwade, Irving T., 1967, "Geologic Quantification: Description-Numbers-Success Ratio," AAPG Bulletin, Vol. 51, p. 1225-1239.
4. Rummerfield, Ben. J. and Morrisey, Norman S., "How to Evaluate Exploration Prospects," World Oil, April 1965, p. 126.
5. Campbell, W. M., "Risk Analysis Overall Chance of Success Related to Number of Ventures," paper given at 36th Annual Fall Meeting of SPE of AIME, October 1961.
6. Kaufman, Gordon M., "Statistical Decision and Related Techniques in Oil and Gas Exploration," 1962 Ford Foundation Doctorial Dissertation Series.

9 Risk Analysis—Applications

Numbers are useful abstractions. They allow us to investigate ideas and concepts otherwise impossible, but they also offer the illusion of precision. Perhaps the last chapter will serve to illustrate that such words as success rate and field size, although often represented by a single number, more accurately describe a range of numbers.

We shall be looking at specific numbers and ranges of possible numbers. The risks of wildcat success will be combined with additional uncertainties into economic risk. Even with compounding uncertainties in these estimates, the goal is to learn more about the investments we make. In the absence of complete knowledge, we describe only what we know; but that is an important start.

Economic risk In this chapter we shall incorporate the knowledge from all previous chapters to arrive at risk-weighted profitability—the end point for any exploratory evaluation. Risk weighting is important to the explorationist because the most probable outcome from his efforts is a dry hole. The definition of a dry hole eludes accurate description. The offshore areas of the U. S. have numerous wells classified as economically dry. They contained oil or gas (often enough to be termed a discovery on land) but in insufficient quantities to pay for platforms and related expenses. A dry hole, then, can be absolutely dry—no oil or gas—or can intimate teasing amounts of non-commercial hydrocarbons.

Begin with hydrocarbon estimates The introduction of field size distributions in Chapter 8 demonstrated the wide range of possible field sizes in a productive geologic trend.

Yet one of the first things we need to know is "What size field?" In previous chapters we have shown a finite cash flow stream for the successful School prospect; but until the outcome (successful or dry) we need to recognize that other possible sizes for the School prospect could have given us other cash flow streams—some larger, some smaller.

The "most likely" size Whenever we settle on a single field size, it can be thought of as our "most likely" case.

By some method we have singled out specific values for the critical parameters of size and multiplied these to get our "best guess" estimate. The most likely case can be the product of:

$$\text{Net pay} \times \text{recovery per ft. of pay} \times \text{productive area}$$

Our "singled out" values would then be a specific pay thickness, recovery factor and areal extent. Suppose for a Typical prospect in a play we assume 1,000 productive acres with 100 feet of sand having a recovery of 1.0 $\overline{\text{M}}$cf/acre-ft. Then:

$$1{,}000 \text{ acres} \times 100 \text{ feet} \times 1.0\,\overline{\text{M}}\text{cf/acre-ft.} = 100 \text{ billion cu. ft.}$$

In Chapter 5 the methods were shown to convert such a reserve into dollars. The School prospect served as the reference field. To refresh your memory, the steps to determine this value were:

1. Calculation of the investment outflow. We listed the exploration and development investments by year and developed their after-tax cash flow.
2. From a production schedule, an inflow of revenue, after tax, was generated.
3. The investment and revenue cash flows were combined for the net cash flow, after tax.
4. Discount factors were applied, in Chapter 6, Table 23, to the annual cash flows to get the DCFR and present value at various discount rates.

Only one of many possible cash flows for the School prospect was calculated. The absolute probability of obtaining our most likely case is near zero. We made judgments for each critical variable to produce our most likely case, but having only one case limits what we should know about an investment. What other possibilities are there? How probable is their outcome?

Payoff table The payoff table answers some of these questions and it represents a good starting point to consider other possible cases. It is from the payoff table that we get expected value.

In considering expected value, we follow a series of steps, each of which requires either a judgment or some information relating to a judgment. These steps are:

1. List the possible outcomes.
2. By each outcome list its probability of occurrence.
3. Show the economic consequence of this occurrence.
4. Multiply the consequence times the probability of occurrence.
5. Sum the products.

We can illustrate this concept by means of the "Typical" prospect previously assigned a reserve size of 100 billion cu. ft. For convenience sake we will assign it the composite risk factor of the School prospect which was 0.29 (Chapter 8); the two outcomes, then, are:

Dry hole	0.71
100 billion cu. ft.	0.29

We must calculate two economic assumptions to go with the two outcomes. First, let's assume an AFIT cost of a dry hole as $1.5 million. Second, we will assume that our calculation of PVP at 10% for the 100 billion cu. ft. was $7.5 million. Following the steps outlined above we have:

TABLE 27
Payoff table
Typical prospect

Outcome	*Probability*	*PVP @ 10% – M$*	*Risk weighted PVP @ 10% – M$*
Dry hole	0.71	−1,500	−1,065
100 billion cu. ft.	0.29	7,500	2,175
Total	1.00		1,110

This arrangement of expected value combines risk and present value and is often referred to as a "payoff" table. The discovery well may or may not be in the zero year. In the case of the School prospect it was not in year zero. For convenience sake, in the illustration of the Typical prospect the zero year is assumed. The AFIT dry hole cost is thus an undiscounted number.

The expected value of the two possible events (shown under the column headed *outcome*) is \$1.11 million if we consider only the well cost as the total investment for the dry case. It is the expected payoff for the situation stated—assuming it could be repeated many times. Remember, a probability has been assigned to the PVP at 10%; we have calculated the PVP of a given reserve size, then estimated its probability of happening. Our expected value reflects our estimate of risk for a specific size field and a dry hole. If our estimates are correct and we can drill many Typical prospects we should average a gain of \$1.11 million per prospect (plus 10% since our \$7.5 million is a PVP at 10%). Finally, there is no implication that this particular value is frequent, probable or even possible. It represents an average risk-weighted value. It has been adjusted by our judgments about the probability of possible outcomes.

Multiple cases Field size distributions help us realize that using just one possible size restricts our view of a prospect's economic potential.

We need a range of cases to depict more accurately the Typical prospect. To keep the illustration simple, a minimum and maximum case will be used. The probability of these additional occurrences must be determined and related with our most likely case to a successful outcome. The sum of their probabilities must add to 0.29. Cash flows for the other cases must be constructed and discounted. The results might be as shown in Table 28.

TABLE 28

Payoff table
Typical prospect

Outcome	*Reserve-Bcf*	*Probability*	*PVP @10%—M\$*	*Risk weighted PVP@10%—M\$*
Dry hole	0	0.71	−1,500	−1,065
Minimum	30	0.09	−100	−9
Most likely	100	0.18	7,500	1,350
Maximum	400	0.02	30,500	610
Total		1.00		886

The addition of the minimum-maximum cases tripled our evaluation work but gave new information.

1. The expected value is now only $886,000. The decrease of $224,000 results from considering two additional reserve sizes, one smaller, one larger than the most likely size.
2. The expected value is very sensitive to the probability assigned the largest reserve. A 0.4 probability for the 400 billion cu. ft. case yields a risk-weighted PVP at 10% of $1.22 million. This small change in probability (if the probability of the most likely size remains unchanged and the probability for the minimum case is reduced to 0.07) will by itself increase the expected value to $1,498,000.
3. The minimum reserve, 30 billion cu. ft., has a present value loss at 10%. From this fact we can estimate the minimum profitable reserve size—it must be slightly larger than 30 billion cu. ft. Keep in mind that this limit is based on our value of money at 10%. If we valued money at 20% a much larger reserve size would be required. If we valued money at only 8% then 25 billion cu. ft. might be an acceptable size.
4. Table 28 also allows the calculations of a risk-weighted reserve as shown in Table 29.

TABLE 29

Risk-weighted reserves

Outcome	*Reserve bcf*	*Probability*	*Risk-weighted reserve bcf*	*Alternate probability*	*Risk-weighted reserve bcf*
Dry	0	0.71	0	0.71	0
Minimum	30	0.09	3	0.06	2
Most likely	100	0.18	18	0.18	18
Maximum	400	0.02	8	0.05	20
Total		1.00	29	1.00	40

The risk-weighted reserve for the Typical prospect is 29 billion cu. ft., even though we still consider the most likely reserve to be 100 billion cu. ft. Do not think of a risk-weighted reserve as in place reserves! Remember that number is a mixture of reserve size possibilities and estimates of the probability of their outcome.

An alternate probability for the minimum and maximum reserve is shown to illustrate the sensitivity to the probability assigned to the larger reserve. The increase to 0.05 for the larger reserve size changes the risk-weighted reserve to 40 BCF! The most likely size is still 100 billion cu. ft., however.

Note that our first set of probabilities yielded the same risk-weighted reserve as the most likely case when viewed as the only answer (100 billion cu. ft. $\times$ 0.29 = 29 billion cu. ft.).

Why risk-weight anything? There is no guarantee that the Typical prospect contains 29 billion cu. ft.; but we expect 100 Typical prospects to yield an average of 29 billion cu. ft. and $1.11 million (PVP at 10%) per prospect.

The value of risk-weighting anything, including field sizes, is to define an unknown quantity with greater accuracy—or to show the limits of one's feelings or judgments about various sizes, probabilities or conditions. There is also the underlying assumption that you have numerous similar opportunities.

A payoff table can also be used to compare the expected value for various choices.[1] Typical examples are the expected value for:

1. Drilling a 100% working interest well versus a farmout of the tract retaining an override.
2. Drilling a 100% well versus some smaller working interest—or a small working interest plus an override.

The procedure is the same. The events and outcomes do not change; only the PVP changes under the various ownerships assumed.

Multiple variables Perhaps you feel more at home estimating the probability of parameters which make up field size rather than field size itself.

Your familiarity here may produce better input data into your risking "model." Usually such a procedure introduces more cases and the need for a computer. We can show how cases quickly multiply when variables are introduced into field size parameters.

We could test a very simple expansion of our problem by assigning two variables to the three parameters which produce reserve size. These parameters are net feet of pay, recovery per acre-foot and productive area. Two values for three parameters produce eight cases—$2 \times 2 \times 2$ or 2^3. Most explorationists define their parameters at least by maximum, most likely and minimum values. Our choice, then, is three values for three parameters making 27 cases or reserve sizes—$3 \times 3 \times 3$ or 3^3. This expansion for the Typical prospect is as follows:

TABLE 30
Typical prospect

Parameter	*Value*	*Probability*
Net pay-ft.	50	0.30
	100	0.50
	190	0.20
Recovery Per Acre-Ft.-$\overline{M}$cf	0.7	0.30
	1.0	0.60
	1.5	0.10
Productive area	800	0.30
	1,000	0.50
	1,400	0.20

The development of the 27 cases is illustrated by the probability tree diagram on Fig. 20. This diagram is the most elaborate and perhaps complex figure in the book. It contains, however, many useful relationships and concepts; and it is worth spending the time to uncover its meaning.

1. First of all it concerns only reserve sizes. However, it includes all possible sizes based on our assumptions.
2. The diagram begins with the decision to drill and shows all possible outcomes. Note the now familiar probabilities of 0.71 dry and 0.29 successful.
3. The first branches show a 71% chance of a dry hole and a 29% chance of a successful well—a gas discovery.
4. Branches from the gas discovery show a 30% chance of 50 ft. of pay, a 50% chance of 100 ft. and a 20% chance of 190 ft.
5. From each pay thickness are probabilities for recovery factors of 30% for 0.7 $\overline{M}$cf per acre-ft., 60% for 1.0 $\overline{M}$cf and 10% for 1.5 $\overline{M}$cf per acre-ft.
6. The final branching from recovery factors shows a 30% probability for a field size of 800 acres, 50% for 1,000 acres and a 20% probability of a field covering 1,400 acres.

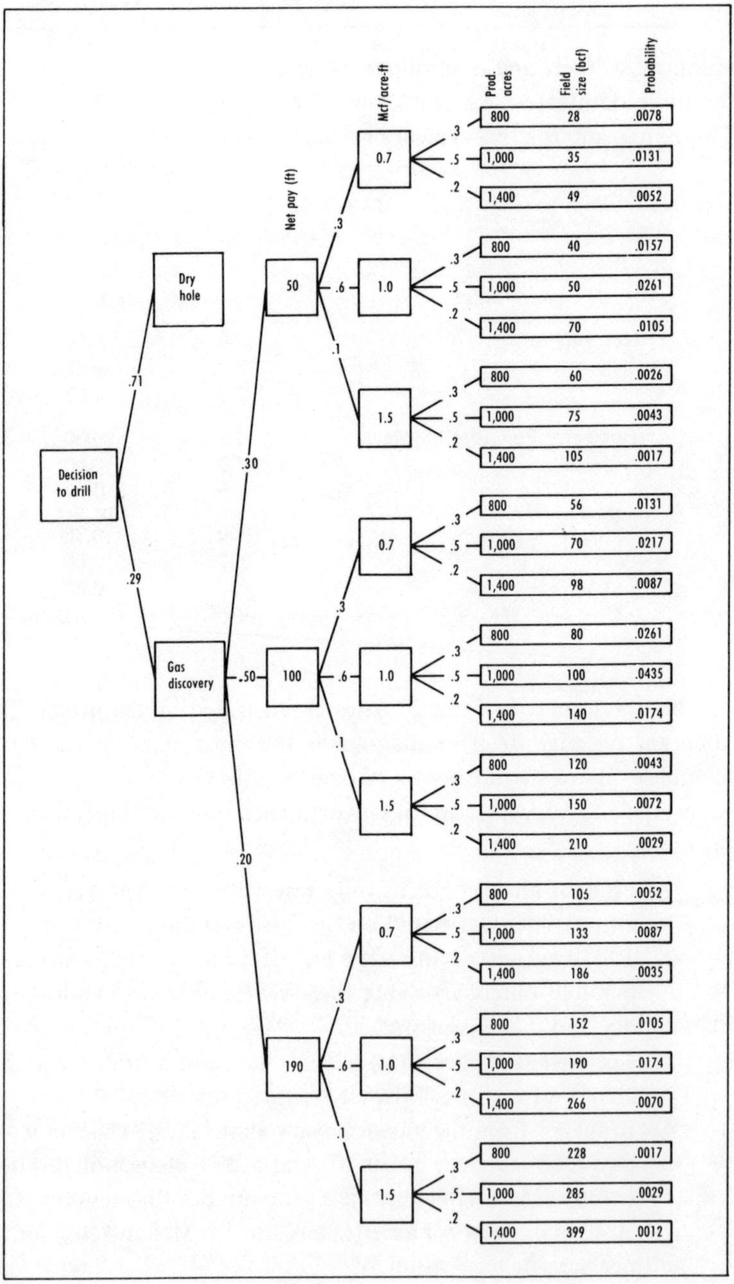

FIG. 20 PROBABILITY TREE DIAGRAM

7. Fig. 20 develops from the three variables for three parameters 27 reserve sizes ranging from 28 billion cu. ft. to 399 billion cu. ft.; and it shows the probability of occurrence of each size.
8. The sum of the 27 probabilities is 0.29—29% the chance of a discovery.
9. Because of the higher probabilities assigned to the middle values of each variable the 100 billion cu. ft. size has the highest probability—4.35%. This number is calculated by multiplying the probabilities of each variable as follows: $0.29 \times 0.50 \times 0.60 \times 0.50 = 0.0435$. The 100 billion cu. ft. size is 36 times more probable than the largest size $(.0435 \div .0012) = 36$.
10. The average of the 27 field sizes is 129 billion cu. ft.
11. The risk-weighted size from the 27 probabilities is 29.25 billion cu. ft.—by design as much as by accident, near the risk-weighted values shown for the single case and the maximum-minimum cases.
12. We show two identical reserve sizes—70 billion cu. ft. Their make up comes from different values for net pay, recovery factor and land area. This coincidence should remind us that various combinations of these three parameters can produce similar reserves but with different probabilities of occurrence.

Things to remember about multiple variables You can quickly grasp the interrelation of our asumptions when viewing the probability estimates shown in Fig. 20.

For example, the initial probability assumption resulted from geologic judgments about wildcat risk—29% chance of a discovery. This number was introduced as a single value, but it also could be a variable. Consider the changes which would result on Fig. 20—all the final probabilities associated with a specific reserve size would change.

The more numerous the assumptions the more numerous the cases. A computer must be used in complex cases, especially if you consider all of the economic variables which will be tacked onto these reserve estimates.

At this point you may want to throw up your hands and return to the simple case of three reserve sizes—maximum, minimum and most likely. Certainly there can be a point of diminishing return in regard to the number of cases. The number of cases required to define a prospect or problem is always a matter of judgment.

One point to remember for variables you want to test is three points define a curve, so test your key variables with three values for plotting purposes. Make sure the values cover as broad a range as you can reasonably envision—then all cases in between can be interpolated from graphs.

Finally since field sizes exhibit a log-normal distribution, prospect estimates logically should do likewise, and they do. See Fig. 21 for the plot of the 27 reserve sizes from Fig. 20. Nature warns us to ponder any field size distribution not log-normal. We can apply the same caution to prospect size estimates.

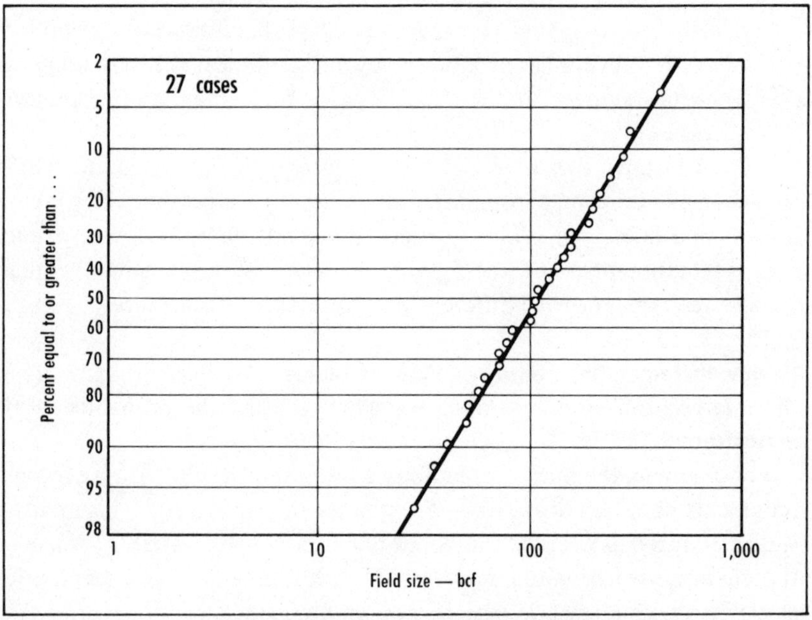

FIG.21 PROSPECT SIZE DISTRIBUTION

Cumulative probability distribution Before we leave our complex situation of 27 cases, one other plot will be shown.

We can generate a cumulative probability distribution which will take into account the probability of a dry hole. Such a curve is shown in Fig. 22. It was constructed by rearranging the cases by size from the largest to the smallest and cumulating the probabilities.

With this curve we can determine the probability of finding reserve sizes throughout the range of the prospect's potential, as estimated from our judgments. The most likely size (or larger) designated previously shows on this curve a chance of 12.5%. The probability of finding only half as much, 50 billion cu. ft., is 25%. Actually the cumulative probability curve demonstrates that we have a most likely range rather than a specific size. That range is the steepest part of the curve.

At the top of the graph a second scale is shown. It is the present value profit associated with these reserve sizes. From this scale the complete range of PVP is shown plus the probability of achieving a given profit. For example, we have a 12.5% chance of making $7.5 million (PVP at 10%) but a slim chance of 2.5% of making $15.0 million. This curve puts the expected value of $1.11 million in a better perspective. The full range of possible profit potential provides a better understanding of the School prospect.

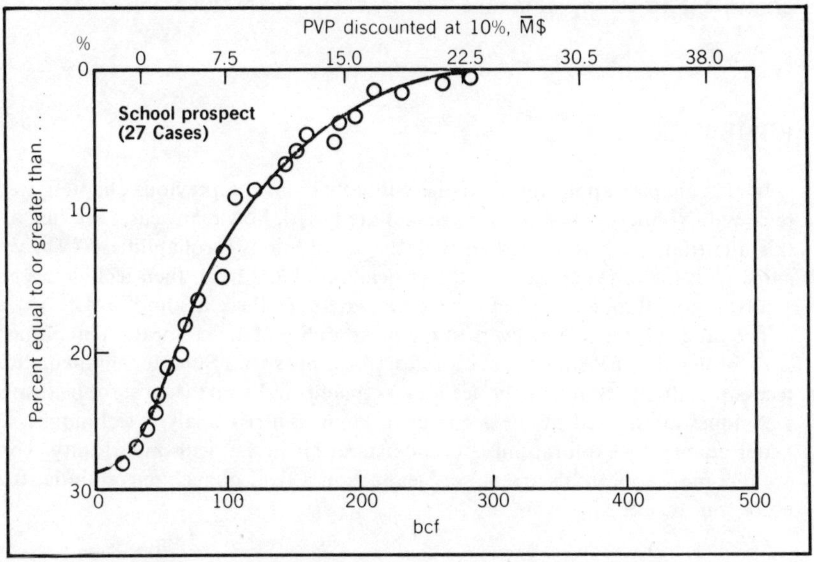

FIG. 22 CUMULATIVE PROBABILITY DISTRIBUTION

Even more complex With computer programs available even more cases could be tried. The formula on p. 134 showed that the number of values raised to the power of the parameters gives the number of cases.

$$(\text{No. of Values})^{\text{No. Parameters}} \qquad \text{Equation 13}$$

Four values for 5 parameters would result in $(4)^5$ cases or 1024 cases. Such a number puts us well beyond the scope of this manual, whose aim is to teach without a computer. You also may be well beyond the scope of your data.

If, however, you desire a number of cases run based on a small amount of input, a technique is available. It is called Monte Carlo simulation.[2,3,4] Those interested in a simple explanation of the method are referred to Appendix D.

Time for reflection Does this chapter demonstrate too many calculations, methods and techniques for a subject with so few facts and such wide ranges of possibilities? Herein lies a significant point. The greater the risks and the more numerous the uncertainties, the less likely that one case or a few cases will adequately define the opportunity. When the range of possibilities is great, we do a disservice by supplying only one answer. It is equivalent to telling someone the next roll of the dice will total seven!

REVIEW

In this chapter applications of the concepts from the previous chapters were reviewed. No matter how many variables are tested, how many cases are run, ask this question, "Are you satisfied with the assumptions (probabilities)?" Do you agree with the range chosen for the critical variables? If so, then techniques are available to tell much of what can be known about these assumptions.

The range of possible answers to the profitability of an exploration investment is so broad that no single case can describe a prospect. Superior mathematical techniques do not eliminate the need for judgment, and even the most sophisticated techniques are limited by the accuracy of input. But risk analysis techniques can better enable the explorationist to understand his investment opportunity. They are one more tool in his arsenal of weapons in a field of such uncertainties that every tool is needed.

PROBLEMS

1. To test your understanding of expected value, solve this problem. (Review the steps in calculating expected value.) Someone offers to pay you $5 for every double six you roll if you give them 20¢ each for all other outcomes. Should you play? Answer by determining the expected value of each roll.

2. Take your favorite prospect. Select some basis for choosing several field sizes. Calculate a risk-weighted field size. Plot your prospect estimates on log-normal probability paper to see if they form a straight line. If you used the method shown in Fig. 20, plot a cumulative probability distribution as in Fig. 22.

BIBLIOGRAPHY

1. Grayson, C. Jackson, Jr., *Decisions Under Uncertainty,* Harvard Business School, Boston, 1960, p. 236–250.
2. Jefferson, James T., *Risk Analysis Improves Pipeline Decisions,* The Oil & Gas Journal, August 31, 1970, p. 80.
3. Smith, Marvin B., *Estimate Reserves By Using Computer Simulation Method,* The Oil & Gas Journal, March 11, 1968, p. 81.
4. Hertz, D. B., *Risk Analysis in Capital Investment,* Harvard Business Review, January–February 1964, p. 95–106.

10 *Setting up an Exploration Evaluation*

The final chapter will deal with the mechanics of setting up an evaluation of an exploratory opportunity. The minimum quantity of data will be reviewed, and its arrangement suggested with some comments on data quality. More than the most likely case will be considered. The most likely case is a useful tool, particularly in engineering economics, where the probabilities have narrower ranges, but in exploratory investments the range of possible answers is usually so broad it lessens the value of a single case.

Almost every exploratory opportunity is unique in certain respects. Therefore, our suggestions on data gathering, data format, evaluation methods and risk analysis do not preclude more specific techniques for any single opportunity. Perhaps the most common decision in exploration involves the prospect evaluation. It shall be used as our starting point. We shall refer to Chapters 5, 6, 7, 8, and 9 when the techniques used have been explained previously.

Gathering the data Our data gathering will begin with the prospect and expand beyond into a play or series of plays.

The gathering of data involves judgments at many points. The most knowledgeable explorationist should assist in the data accumulation to insure quality and to enable him to influence judgments requiring exploratory experience. Place the same value on the input that will be placed on the output—keeping in mind, of course, that some variables are more critical than others.

FIG. 23 EXPLORATION PROSPECT EVALUATION

Prospect name ______________ Location ______________ Date ___________

BASIC EXPLORATION

Total acres in prospect ________________; your net acres ________________

Lease bonus—$/acre ________; annual rental—$/acre ________; royalty _____

Geological effort—months ______; geophysical effort—months & $ __________

Other exploration effort—$ ____

(Test wells, core drilling, overhead, etc.)

Composite risk factor (or success rate)—% _________________

WELL DATA

No. wildcats to evaluate prospect ______; depth ________; your W.I. ________

Successful wildcat—$: tangible ___________; intangible ___________

Dry wildcat—$ ___________; well spacing—acres ________

Successful development well—$: tangible ___________; intangible ________

Dry development well—$ __________; development drilling success rate—% _____________

Operating costs—$/completion/year ______________

TIMING—Calendar Year(s)

Year of sale of production ______________

Leasing ________; Geology & geophysics ________; drill wildcat (s) ________

Development well schedule—wells _____ _____ _____ _____ _______

year _____ _____ _____ _____ _______

RESERVES (If more than one zone, use separate sheet.)

Pay zone name _____ Oil or gas ___ (If both: ratio of gas cap to oil zone___)

(If gas: inert gas % ___________; condensate—bbl/$\overline{\text{M}}$cf ___________)

Productive acres: total ________; your acreage ___________; depth ________

Average net pay—ft. ___________; recoverable reserves/acre-ft. ____________

Allowable ___________; secondary recovery ___________

REQUEST

Type evaluation desired (variables to be tested) ___________________________

Remarks ____

Date needed ____________________ Requested by ____________________

On Fig. 23, the basic data needed for a prospect evaluation are shown. The data are segregated into four categories each of which will be briefly discussed.

Basic exploration The beginning statistics deal with acreage. How large is the prospect in acres?

How much do you own? What was paid per acre, or is expected to be paid on acreage as yet not acquired? List the rental conditions and royalty considerations. All other exploratory data, the physical units and their corresponding costs, should be tabulated—and related to time, i.e., in what year was the money spent. Some years may be future ones.

If you have not calculated a composite risk factor (and many analysts do not), what success rate (in per cent) is anticipated for this trend? Is it applicable to this prospect?

Remember that some of your data will represent capitalized costs and other data expensed items.

Well data Statistics about both wildcat and development wells need documentation.

Will more than one wildcat be needed to test the prospect? How deep? What is your working interest? Separate all well cost estimates into tangible and intangible segments. If a dry wildcat will cost less than a successful one (and it usually does), show its cost. What well spacing do you expect? How many successful and dry development wells will it take to develop the prospect and what is their cost? The direct operating costs associated with production should be recorded as well as indirect costs and overhead gleaned from appropriate accounting records.

Timing Review Tables 11 and 12 (Chapter 5) for a reference on timing.

Because we will discount future expenditures and revenue, timing is important. Consider carefully the scheduling of exploration expenditures and the drilling of development wells. The normal temptation in prospect evaluations is to over optimize the scheduling of exploratory expenditures and the rate of field development. Events can and often do alter our rate of exploration and field development. To expect everything to come off smoothly ignores a lot of painful history. Be realistic in timing. If in doubt make timing one of your variables.

Reserves In view of all that's been said, the reserve data requested may seem skimpy.

It is. What is shown is absolutely minimal and would apply only for a most likely reserve estimate. Are you exploring in a mature trend? Do you have access to a field size distribution; if so, it is your starting point. Whatever reserve sizes you select, the field sizes should fit a log-normal distribution as outlined in Chapter 8. In prorated states the allowable is necessary information. For gas wells the rate of take (DCQ) must be estimated. Allowables and takes must often be related to knowledge about reservoir characteristics. Is the formation of such low permeability that deliverability is limited? Special field problems must also be noted; examples would be: high per cent of inert gas, CO_2 perhaps; presence of H_2S; remoteness from market; unusual shrinkage, and so forth. The data category on reserves could obviously be expanded; just how much will depend a lot on your data base and whether you have access to a computer.

Defining the request The type of analysis needed must be carefully considered. Both the requester and analyst should participate in defining the request. In areas where a single factor (reserves, bonus per acre, success rate, etc.) cannot be stipulated, some sensitivity testing should be undertaken. Begin by asking, "What do we want to know about this prospect?"

Are you trying to determine what bonus you can pay? If so, set up several cases to check various bonus prices which can be paid. Other variables can be tested by varying the reserve size, success rate, etc. Are risks quantified or at least tested for their impact on potential profits?

The definition of a request is a key step. When the important factors are carefully reviewed, the result is a well-defined evaluation and many fewer recalculations.

Sensitivity testing Some comments have already been made on sensitivity tests. In general you should test the sensitivity of any factor recognized as critical. A critical factor is one in which a possible change makes a significant change in profits or guideline yardsticks.

We can demonstrate a few factors using the School prospect. In Table 31 the sensitivity of several changes on yardsticks of the School prospect is noted:

TABLE 31

Sensitivity checks—School prospect

Case No.	*Assumption*	*AVP M$*	*PV @ 10% – M$*	*DCFR %*
1	Base case	5,200	1,583	22.2
2	Doubling exploration expenditures	3,300	−44	9.8
3	Doubling production investment	3,300	659	12.1
4	Two-year delay of revenue	5,200	804	14.5

From these simple illustrations, the important variables can be inferred. Case 1 is the base case used in prior chapters for the School prospect.

One of the critical variables for the explorationist is the amount of exploratory expenditures. Case 2 in Table 31 shows the results of a doubling of exploration expenditures. (You can also think of this as cutting your success rate in half.) Spending twice the scheduled amount in finding the School prospect reduces the total money received (AVP) to $3,300,000. It eliminates the present value profit, if we value our money at 10%—which means the DCFR is less than 10% or 9.8%.

Notice the effect of doubling the development expenditures, Case 3. The AVP is the *same* as doubling exploration expenditures; but because our development investments take place later in the cash flow stream—and are thus discounted more—time value yardsticks are affected less. There is still a present value profit at 10% and the DCFR is 12.1%.

In Case 4 the effect of a two-year delay in revenue is shown. The AVP is the same as the base case. We get our money later, however, and the time value yardsticks reflect the loss. The PVP at 10% is reduced from $1,583,000 to $804,000; and the DCFR is decreased to 14.5% from 22.2%.

Complications of new bidding legislation The state of Alaska and the Federal government have recently passed legislation which outlines several new bidding methods and involves new types of evaluation calculations.

In addition to bidding a cash bonus at a fixed royalty the new legislation provides for:

1. Royalty as the bid variable
2. Net profits bidding
3. Repayment of some expenditures, from a bonus bid, related to exploration.
4. The amendments to the Federal OCS Lands Act allow a sliding scale royalty and other alternate bidding methods.

These new bid variables are akin to sensitivity testing in that they require an understanding of each related to a standard profit yardstick. Certain of these bids allow a high value for the bid variable to achieve an acceptable DCFR, but very little actual value profit. They, thus, require new complex calculations and new dimensions in yardstick values.

In general, bid variables which require low cash input may also produce low cash output. Although risk can be de-emphasized by such bid variables, the profit reduction may overwhelm the risk reduction to the point of limited interest in the sale.

The new variables in sealed bid sales in Alaska and in Federal sales on the Outer Continental Shelf will generate a greater volume of analytical paperwork. The benefit, if any, to the consumer has yet to be shown.

A new admonition for the analyst is to keep up with new and potential legislation to see what effect it has on your bidding concepts, your standard computer programs and your value concepts.

Inflation—a form of sensitivity testing In any cash flow analysis the question arises as to how to treat inflation. In times of modest inflation, no more than 2 to 4 percent per year, inflation can almost be ignored. Why? It gets "swamped" by the more critical parameters such as risk or reserve size. However, at higher rates of inflation, greater than 6 percent, it should be considered.

Opinions on how to handle inflation differ. They center around three approaches as follows:

1. *Constant dollar method.* In the constant dollar approach the projections of the various components of cash flow are expressed as dollars of the year in which the evaluation or analysis is made. This method assumes that prices and costs will inflate at the same rate and thus offset each other—at least approximately. In effect, you are ignoring inflation in using this method, if your assumption is not valid.

2. *Component inflation method.* In the component inflation method the inflation rate for each cost or price component in the cash flow stream is forecast independently. For example, you may feel wages will increase at 6% per year during the life of your investment but that material cost will increase at 10% per year. If so, you take the wage-related costs and increase them at 6% per year. Material costs are increased at the higher rate. Using this method you apply your best knowledge to each component of the cash flow stream. Thus wellhead prices for oil and gas, drilling costs, lease and well expense, leasing costs, etc. can be forecast based on recent experience or reasonable future expectations.

3. *The single inflation rate method.* This method applies a single rate of inflation to all cost (or price) ingredients. The assumption precludes individual component rates. It would apply where you feel data insufficient to develop realistic or accurate rates of inflation for individual components of cash flow.

Some analysts go one step further. They say, in effect: "All this inflation forecasting is fine—but in 1990 when the cash flow stream returns $100,000, that sum will not buy as much as $100,000 today!" Thus a "deflator" is advocated to correct future dollars to what they would buy today. Even this criticism does not apply in all cases. It would probably be more appropriate for drilling costs, but certainly not so for rapidly changing technology—small electronic calculators being the prime example of a *decreasing* cost item, not an inflating one.

So what do you do? None of these methods promises absolute accuracy. All involve key assumptions about the future. If you think you have good information on a few parts of the cash flow stream, you may favor the component method. You can throw up your hands and inflate not at all, or use a single escalation factor. You have to choose according to what you believe will provide the most realistic answer—the one which will enable you to best optimize your investment policy for the long pull. However, if the rate of inflation approaches double digits you need to take it into account.

The critical variables For most exploration investments, the significant variables are:

1. Risk.
2. Amount of exploratory expenditures.
3. Field size—reserves per well.
4. Deliverability (especially for gas discoveries).
5. Delay in revenue.
6. Wellhead price.

Each of these variables will be discussed briefly. No specifics will be given because of the unique characteristics of each prospect, plus the special conditions associated with the hostile environments of Alaska and offshore. In offshore environments where development, facility and producing costs are large, these costs become important economic parameters. The six variables we have listed would be critical to any exploratory investment whether onshore or offshore.

Risk The investment cash flow in Table 11 for the School prospect assumes at least one, possibly two, dry wildcats. In making this assumption, we are departing a bit from a normal one-well-on-a-structure evaluation. It is done for illustration purposes. The methodology recommended here is just as applicable if only one well per prospect is assumed. Assuming that two dry wells were drilled, then the play resulting in the successful School prospect had a success rate of 33%, slightly above the calculated wildcat risk. In Table 31, Case 2, doubling the exploratory expenditures without doubling the revenue is about equivalent to halving the success rate—say to 17%. We see then that the DCFR of the School prospect is very sensitive to "risk" changes. Furthermore, to achieve a 10% DCFR will require almost a 20% success rate, assuming we have a large number of similar plays

Whatever basic data are available to measure risk, it is one of the key variables. Consider it carefully and check the effect of various risk values on your prospect's cash flow. One simple method to measure the effect of risk is a plot of PVP at your desired minimum discount rate versus the number of dry prospects per successful prospect. The use of dry prospects per successful wildcat produces a straight line on your graph so only two points are needed. With a computer three points are just as easily obtained as two and will verify your line.

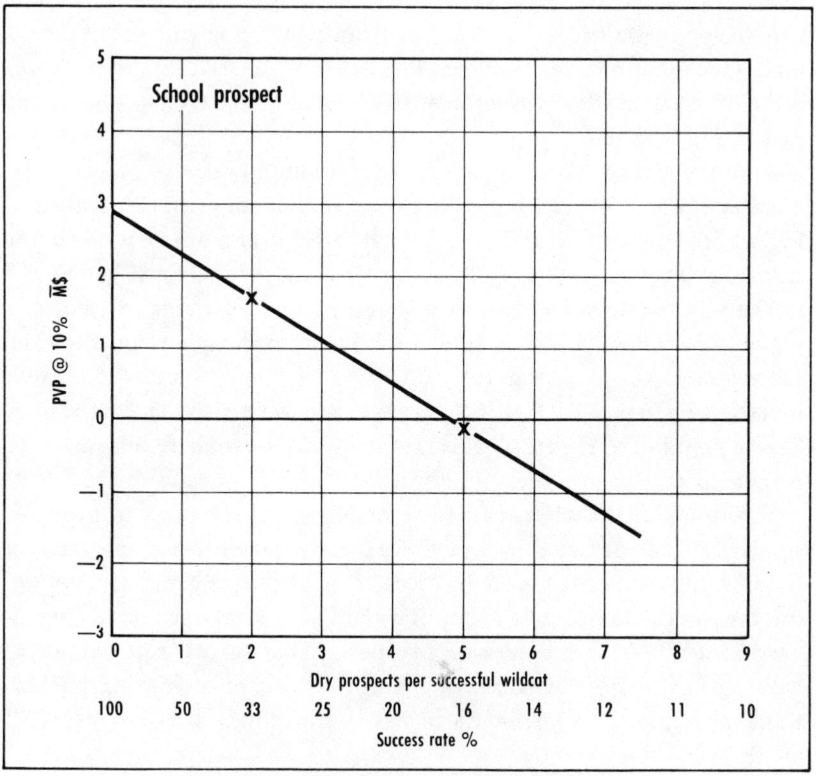

FIG. 24 MINIMUM SUCCESS RATE TO OBTAIN 10% DCFR

Fig. 24 has such a plot for the School prospect. The two points for the graph are PVP's for the base Case 1, and the doubling of exploration expense—representing 2 dry prospects and 5 dry prospects per producer, respectively. The straight line was extended to the Y axis and to the zero PVP line. The present value profit at 10% becomes zero at 4.85 dry prospects per success—so the limiting success rate is:

$$\frac{1}{4.85+1} \times 100 = 17.1\%$$

(The successful wildcat divided by the dry wildcats plus the successful one times 100 equals the success rate.) This per cent is somewhat theoretical. You might not have enough prospects in the play (like the School prospect) to test such a success rate. The value lies in the limit. Suppose the minimum permissible success rate is 2 or 3 per cent and that your experience in the trend has been substantially higher (say 25 or 33%): then you can feel reasonably confident that your prospect is not limited by economic yardsticks or the other assumptions from which you calculated the 2% limit. One other point comes from Fig. 24. The intersection with the Y axis is the PVP with no dry prospects, or 100% success. The PVP at 10% for this case is $2,900,000.

Another graphic form to obtain the minimum success rate involves a graphic plot of the risk-weighting of an individual prospect. Shown on Fig. 25 it requires the AFIT cost (PVP at 10%) of a dry venture (in this case −$200,000) and the results of a successful prospect—$620,000.

The plot for Bearcreek prospect shows that a 25% chance of success is required to achieve a zero PVP at 10%. It then is the minimum allowable success rate. You can see graphically that any smaller success rate would necessitate a higher PVP at 10% (in effect a larger prospect). On the other hand a guaranteed higher success rate—say 50%—would require less PVP from a successful prospect—in effect a smaller prospect.

Variations of this chart can serve as graphical solutions to problems similar to those depicted by a payoff table. Suppose for example that you thought the chances of a 25% success rate were questionable. Suppose further you calculate that a farmout with favorable retained royalty would yield $300,000 from a successful prospect. What success rate would you have to have before it is worth taking the risk of a working interest? Remember that in a farmout you have zero money at risk—zero PVP at 10%.

The solution is shown in Fig. 26. The farmout line crosses the working interest line at 40%. Thus, you would need a success rate greater than 40%

for the working interest venture to give you a better profit than the farmout. You are assuming that your probability risking is accurate, of course. However, if the 25% estimate leaves you uncomfortable the answer is—farmout.

Graphic solutions sometimes offer simple pictorial explanations for analytical problems.

Exploration expenditures The comparison of Case 2 to Case 1 on Table 31 demonstrates the significance of exploratory expenditures to time value yardsticks.

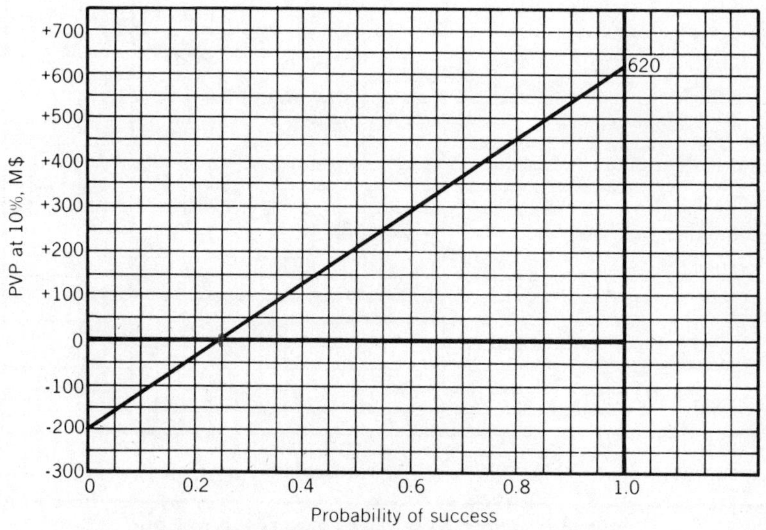

FIG. 25 MINIMUM SUCCESS RATE AT 10% PVP
BEARCREEK PROSPECT

Remember that exploratory investments start at the front end of the cash flow stream and are discounted very little. As such a dollar of exploratory expenditures is more significant than a dollar of development expenditures, the latter is farther down the cash flow stream. Comparing Case 3 to Case 2 in Table 31 shows this relationship clearly. The AVP is the same for both cases, but because the development expenditures are later (discounted more) a significant difference exists in their effect on time value yardsticks.

Field size and reserves per well The total quantity of reserves found is important, particularly if the exploratory expenditures are large.

The recovery per well is equally important. Once sufficient reserves are produced to overcome the effect of exploratory cost, the DCFR will approach that of the average development well. Obviously the DCFR for a prospect cannot exceed that of the average of the development wells. This last statement has special meaning for high cost areas. You may have a trillion cu.-ft.-gas field, but if it is deep (say 25,000 ft.) and you can only produce 2 billion cu. ft. per well, it is uneconomic. The reserves per well are important.

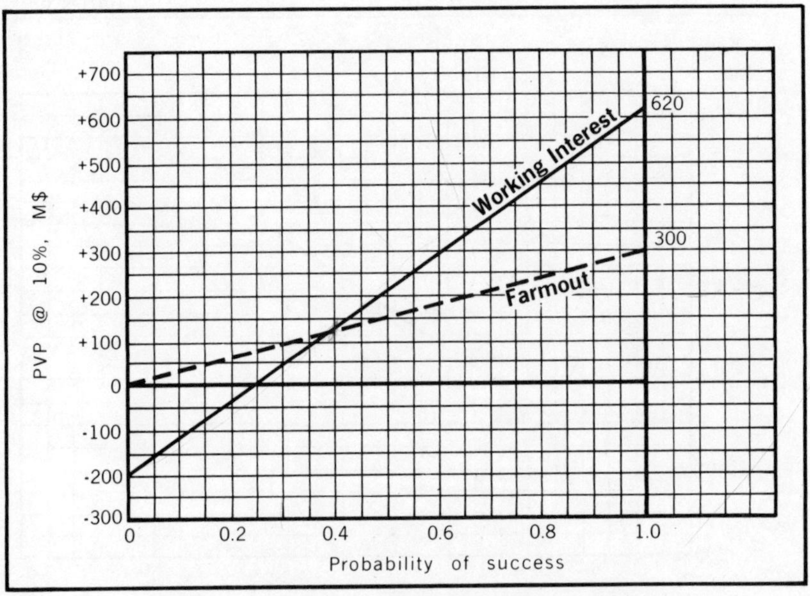

FIG. 26 MINIMUM SUCCESS RATE AT 10% PVP
BEARCREEK PROSPECT

Deliverability Closely related to recovery per well is the actual rate of flow from an oil or gas well. By now you realize that any factor which lengthens payout lowers the DCFR. One way to lengthen payout is to have a well with a very low production rate. Since most oil wells are shallow with attendant lower investments, deliverability is more critical with deep gas wells. A deep gas well needs a high flow rate for profitability. If you face low permeability at depth and thus low producing rates, you have a very critical parameter.

Delay in revenue Case 4, Table 31, illustrates the effect of delay in revenue on present value yardsticks. Delay can come about from mechanical well problems, a slow development rate for the field, restricted market outlets, awaiting construction of complex lease facilities or a pipeline and so forth. The more delay the longer the payout (at least on the exploratory investment) and the lower the DCFR.

Delay is one factor over which the explorationist often has no control. Nevertheless, if in doubt check the sensitivity of yardsticks to possible delays. Not to do so would be to fail to describe the uncertainties of the opportunity.

Needless to say, you can do a lot more checking of sensitivities with a computer available. With practice and experience, however, one can spot almost by inspection the variables which need testing for sensitivity. Some factors not mentioned would be critical in different evaluations. A partial analysis of a downstream investment might have costs critical to profits. In exploratory investments, the costs of exploration precede other investments by such a time margin that they are much more critical. For this reason the experienced analyst spends more time and care on the exploratory phase of his investment stream than on other downstream investments. The importance of timing on yardsticks requires it!

A delay in both revenue and investment *of the same time period* will not change the DCFR. However, it will lower the PVP for all discount rates below the DCFR. In the case of the School prospect, for example, the effects of delay (both revenue and investment, remember) are:

	Delay—Years		
	0	*5*	*10*
AVP—M$	5,200	5,200	5,200
Discounted Profit—M$			
5%	2,950	2,311	1,811
10%	1,583	985	610
20%	189	80	29

The graphic expression in loss of discounted profit is shown on Fig. 27. The lower PVP with delay clearly shows one of the weaknesses of using only DCFR as *the* yardstick. As we have said, there is no *perfect* single yardstick.

Price In the first edition of "Exploration Economics" price was not listed as a critical variable. It has become a most critical variable since the large

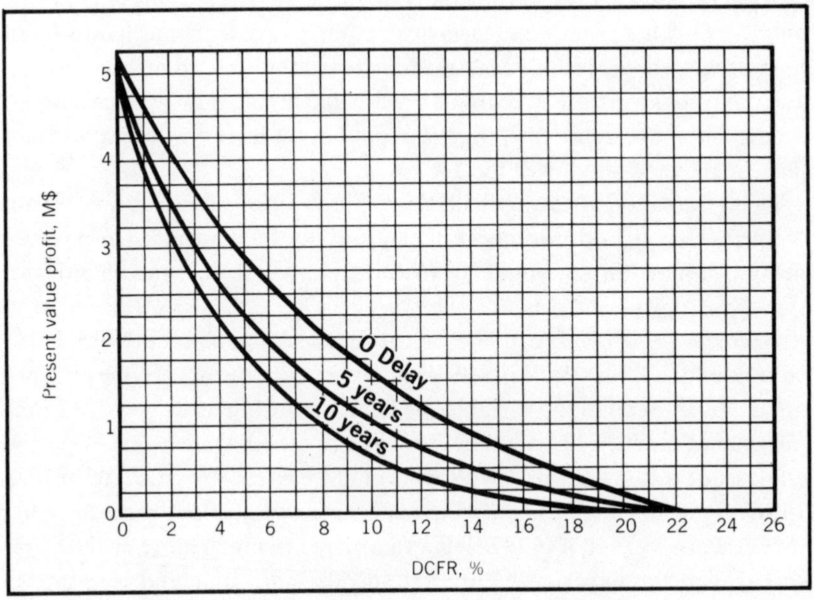

FIG. 27 THE EFFECT OF DELAY ON PROFITABILITY

increases in inflation since 1973. Prices regulated at low values often preclude further field development for both "old" oil and interstate natural gas. Furthermore, inflation is always two "jumps" ahead of regulatory bodies' ability to react.

The new energy bill (1978) includes prices for natural gas which add mostly confusion. There are from 15 to 20 categories of gas, each with its own price schedule. Furthermore, the act is not deregulation as touted—but more regulation since for the first time *intrastate* gas would be regulated. Only the government can offer more control and call it decontrol!

Price is a key variable. If in doubt about a price ruling test your prospect at the worst (lowest) price schedule. Would you still be willing to make the investment? Don't forget a natural gas or oil price ruling can be given *after* you have already made your investment. So test the sensitivity to various price assumptions. Then make your best judgment.

Play vs. basin risk If you work for a large company, a plot similar to Fig. 24 can be useful. The successful exploration programs must carry the unsuccessful ones—a fact overlooked by many of our critics. If you wish to check the "carrying" ability of a successful play, plot PVP, at your guideline discount rate, versus dry plays per successful one. The number of dry plays your successful one could carry occurs where the line crosses zero PVP.

The economics of new basins One of the most baffling problems facing the explorationist deals with the economic results from exploration in virgin basins.

Where no oil or gas has been found, the facts are few. This type of evaluation is more geologic than economic. The assumptions made concerning the presence or absence of hydrocarbons, depth and type of reserve and field size will normally outweigh all other input factors in the effect on economic yardsticks.

What, then, can the analyst contribute that will assist in the decision? He can show the results of a proposed exploration program and determine the magnitude of reserves required to make a profit from these investments. If the basic assumptions do not produce desired yardstick levels they can be re-examined. Changes may produce different yardstick levels. If no changes are felt possible or necessary, then poor economic results suggest caution. A different level of expenditures can be planned.

In summary, then, an evaluation of a program in a virgin basin can only show the economic results of assumed reserves and risks and proposed expenditures. If the field size and number to support the program appear reasonable, proceed. If your most reasonable assumptions do not produce profitable results, the warning flag is out.

Another sensitivity check If you wish to illustrate sensitivities graphically, a common means is by use of a spider diagram. These diagrams have been shown in evaluation literature for many years.

On Figure 28 a sample of such a sensitivity chart is plotted. The ordinate is present value profit (PVP) at the desired discount rate. The abscissa shows the base case at 1.0 and deviations from the base case. To the right parameter values are greater than the base case, to the left smaller.

The base case parameters are:

30 BCF Prospect
1.0 $\overline{\text{M}}$cf/D Initial Potential (IP)
1/6 Royalty
50% Success Rate
$25 Per Acre Bonus

Each parameter must be varied at least twice to get a curve—one value above and one below the base case.

Sensitivities are indicated by the curves drawn on the chart. All curves pass through the ratio of 1.0 since it is the starting point for either plus or minus deviations. Thus Figure 28 shows how field size, initial potential, success rate, bonus and royalty vary from the base case but only *if all other values remain constant.* The last phrase is important. The curve indicated for initial potential shows changes in PVP at 10% for that parameter alone, with all other parameters remaining at the base case value. Therefore, one of the limitations of spider diagrams is that they test only one parameter at a time.

There are other parameters not tested. Examples are well cost, operating cost, wellhead price, etc. The choice of which parameters to evaluate must come from the analyst. Those chosen should be only the factors in which small variations make large changes in profits.

Spider diagrams allow useful comparisons. For example, suppose you had to give a ⅓ royalty instead of ⅙. Looking toward the right side of the chart one-third royalty is equal to a ratio of 2.0 relative to the base case value. You can see that the PVP at 10% is a loss of almost $3 million if the royalty goes to one-third. Note also that to compensate for such a high royalty it would require 1.6 times the base case initial potential or 1.5 times the field size. In this sense, spider diagrams can compare two variables simultaneously showing what positive effect could overcome the negative effect on profits caused by other factors. A spider diagram is a sort of trade-off chart because of the ability to compare changing parameters to a base case.

Choosing the base case is important. Normally it should be at the lower end of field size so as to check marginality. A base case could be chosen so that the PVP at 10% would be zero.

In the particular instance shown, the bonus per acre, in the range expected, is not a critical factor and could be eliminated from the graph. The greater the slope of a curve the more significant the variable is to the project. Again, the significance of the base case is illustrated. A base case of $500 per acre would generate a bonus curve with much greater slope.

Spider diagrams demonstrate the profitable limits of parameters when the PVP becomes zero. In the example shown in Figure 28 the base case was chosen to coincide with zero PVP at 10%. Thus lines below zero denote negative effects on profits, those above positive effects. For example, lower bonus and royalty produce greater profits than the base case. Conversely, higher bonus and royalty reduce profits. Larger field size, IP, and higher success rates increase present value profits. Lower values for each parameter reduce profits.

How can you use this diagram as a trade-off chart? You use it in this manner—balancing a negative effect on profits to an equal positive effect. For example, what does it take to offset an increase in royalty to one-fourth (a ratio of 1.5 relative to the base case)? At 1.5, the negative effect of increasing royalty to one-fourth is about \$1.5 million (PVP at 10%). To offset this loss requires an increase in IP to 1.25 times the base or 1.25 million cubic feet per day. Therefore, if the sum of the negative and positive effects equals zero, the changes in the two parameters equal a trade-off.

Spider diagrams have limitations:

1. They are related to a single case from which each variable is increased or decreased in value.
2. You must choose the right parameters to vary.
3. The number of cases needed to build a spider diagram usually requires access to a computer—not a serious limitation to most explorationists.

Spider diagrams do not have to be constructed so that the base case has a present value profit of zero. It is advisable that a negative PVP can be shown for each parameter to enable trade-off comparisons. A base case near zero PVP would suffice almost as well as the example shown.

Spider diagrams test sensitivity in a unique way and allow a visual representation of the trade-offs between parameters. If you have access to a computer you may find this representation valuable. You can always draw more than one spider diagram if you wish to test more than one base case.

Using FSD curves when discovery size is decreasing Many economists and mathematicians in dealing with a finite number of fields would prefer analyses which recognize "sampling without replacement." In other words, does your analysis take into account the fact that each new field found subtracts one from the total available to be found? The math for sampling without replacement is not unduly complex, but one can handle the problem with adequate accuracy by means of shifting field size distributions (FSD).

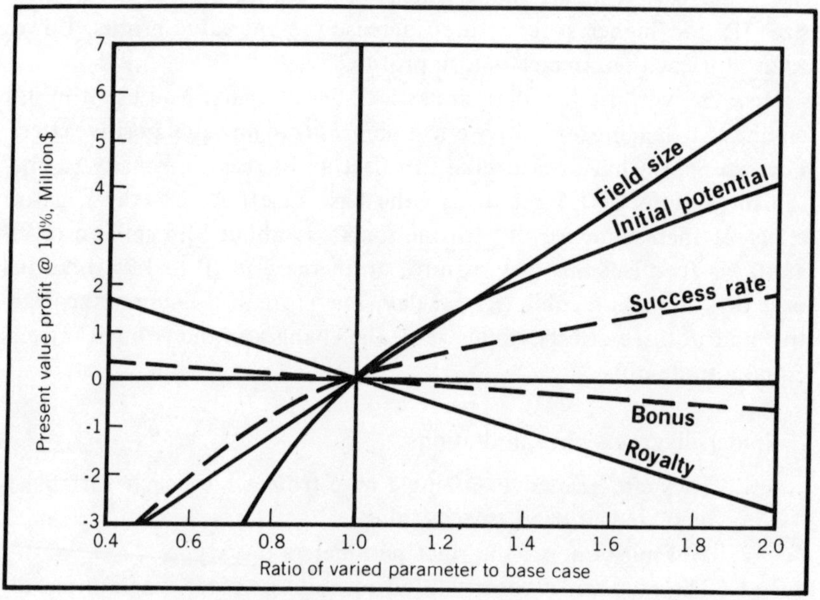

FIG. 28 PROFIT VS. VARIED PARAMETER

On Figure 29 several FSD's are shown. Each is designated by a lower case letter. Curve "b" is the entire population of fields found to date. It represents all fields discovered in the Great Gulch Basin from 1942 to 1978. Curve "a" represents the FSD for fields found the first eleven years of basin life—1950 to 1960.

Note the larger size of fields in the beginning. The median (50th percentile point) for curve "a" is 20 million barrels. For the entire history of the basin (curve b, 1950 to 1978) the median size is only 10 million barrels. Obviously the more recent discoveries have been smaller. Sure enough curve "c" shows that for 1965–1978 the median had slipped to 5.2 million barrels. Discoveries are getting smaller in Great Gulch Basin and the analyst must recognize this change. How? The way to do this is to estimate curve "d"—a projection of a future distribution covering the period 1978–1985.

Curve "d" will not be absolutely accurate, but it does move in the right direction. You can estimate "d" by plotting the median size versus time (actually the midpoint of a time period) then projecting the median size into the future. One thing you do know, this method is better than using curve "c" or "b" or "a"—if you are looking for the same type of prospect or

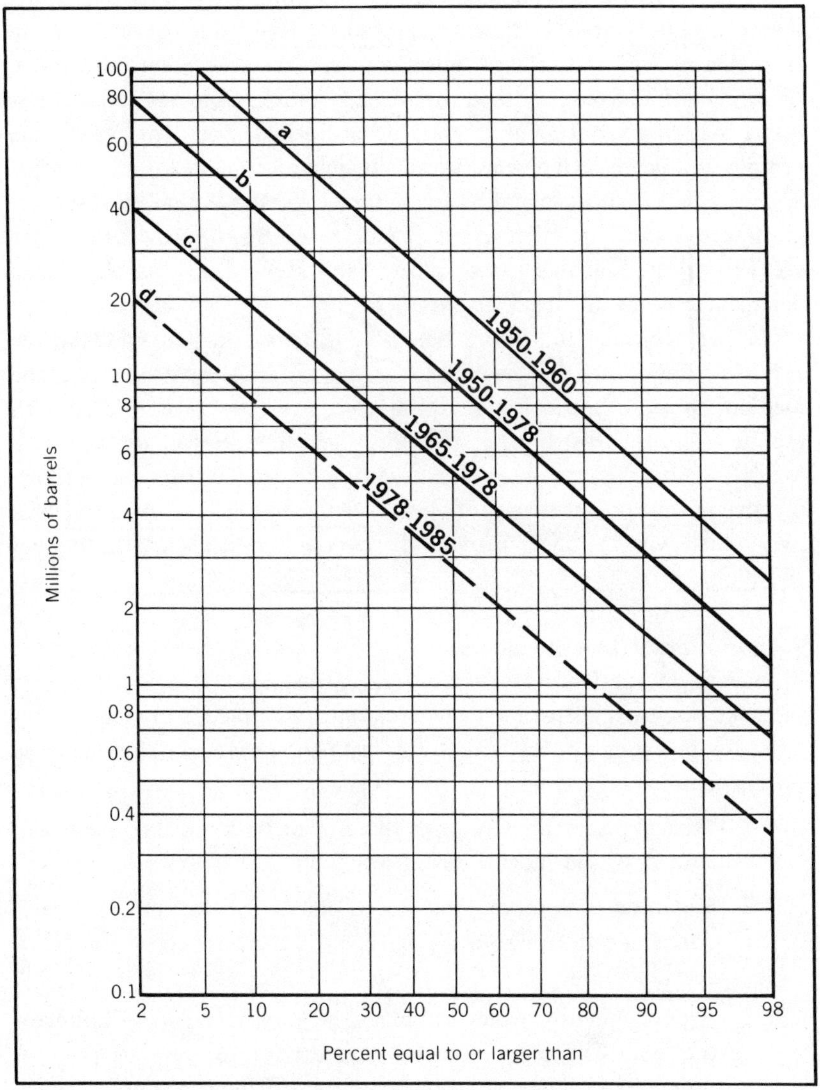

FIG. 29 FIELD SIZE DISTRIBUTIONS CHANGE WITH TIME GREAT GULCH BASIN

objective. New technology could, of course, cause a new wave of larger discoveries from deeper depths or a newly recognized objective. However, this method is to be used when you have assumed you are essentially continuing an existing play.

Before we leave this rough method for sampling without replacement a few comments. Curve "b" represents the entire basin; it is all the history on sizes. Curves "a" and "c" are subsets of "b." They represent portions of "b" and would have fewer fields involved. As such, from a practical viewpoint they might not yield as good a log-normal plot. However, if the sample is large enough, even subsets will yield reasonable log-normal fits.

Experience in most inland basins in the U. S. shows that most subsets are approximately parallel to the full data set. So, in projecting a subset the projected median can be plotted and a line drawn parallel to the basic data set, in this instance curve "b."

Having projected the future field size distribution and knowing the average success rate you analyze a sample of fields (or if only one, the median) to see if you can make money from the fields associated with the new (future) distribution. By this method you will answer in a logical manner, the question of can you make money from the decreasing size of future discoveries. You also determine when to get out of a play which will no longer provide discoveries of sufficient size to yield acceptable profits.

Review of evaluation procedures At this point we have completed the evaluation of the School prospect, having carried it from a lead to a discovery. As a final review for the chapter, a sequence of steps is outlined for prospect or play evaluations with the appropriate reference to prior chapters.

1. Gather pertinent data with the aim of building the investment cash flow and income cash flow.
2. Separate exploratory and development investments into categories shown in Table 11. Then build the investment cash flow stream—Table 12.
3. From regulatory rules establish the flow rate per well and for the field. Consult a reservoir engineer for advice on the proper flow based on reservoir characteristics. From crude or gas price, calculate annual revenue and following the steps in Tables 13 and 14 calculate the income cash flow.
4. Combine investment and income cash flows to get net cash flow (AFIT).

5. Adjust cash flow for sensitivity tests desired and discount base case and adjusted cash flows.
6. Consider risk by the payoff table, success rate or whatever method best describes your investment opportunity.
7. Calculate yardsticks as outlined in Chapter 7. Show from the sensitivity and risk calculations the change in yardsticks caused by critical parameters.
8. If satisfied with the economic and geologic description of the opportunity make a recommendation or decision.

A word of caution. When dealing in prospect evaluation, the decision point, year zero, occurs after some funds have been spent. Frequently these prior expenditures are excluded because the decision point is *now*. However, each prospect must bear its proportionate share of all exploratory costs—otherwise you would never fully account for all costs. One can raise the guideline DCFR a few points higher to compensate; or if the average prospect shows a DCFR well above the minimum guideline, ignore prior costs but remember that the actual rate of return is less than stated.

EPILOGUE

The time has come to ascertain if the goals at the outset have been reached. The reader is the best judge. In a primer, one must deal, purposely, somewhat behind the state of the art. The principles and procedures included are considered necessary as the footstool to further investigation.

No evaluation technique has been intended to be shown as absolute or final. History demonstrates continuous evolution of techniques and today's techniques are part of this continuing evolution. The literature of today reflects efforts toward further refinement of existing techniques as well as a few new ideas; and the literature is always behind the state of the art, particularly that from the larger companies.

No apology is given to the expert for the generalizations necessary. In the beginning one must state positively a few ideas which should later be qualified to some extent. This book is no exception.

If a new idea or two from this book helps you understand the next evaluation, or if you feel able to discuss exploration investment evaluations more easily and intelligently, then the efforts have been worthwhile and the goals set forth have been achieved.

PROBLEMS

1. Using the cash flow streams for the School prospect from Chapter 5, prove that doubling the exploratory expenditures eliminates the PVP @ 10%.
2. Prove that doubling the development investment has less effect than a corresponding change in exploratory investment.
3. Illustrate the effect of the two-year delay in revenue for the School prospect.
4. Illustrate a graphic solution for DCFR from discounted present values in Problem 2, Case 3.

Appendix A

TABLE 1
Summary of allowables
State of Texas

Depth	*1947* Depth yardstick b/d				*1965* Depth yardstick b/d					NPX	Marg. all.
	10 Ac.	*20 Ac.*	*40 Ac.*	*80 Ac.*	*10 Ac.*	*20 Ac.*	*40 Ac.*	*80 Ac.*	*160 Ac.*	*b/d*	*b/d*
0– 1,000	18	28	—	—	21	39	74	129	238	20	10
1,000– 1,500	27	37	57	97	21	39	74	129	238	40	10
1,500– 2,000	36	46	66	106	21	39	74	129	238	40	10
2,000– 3,000	45	55	75	115	22	41	78	135	249	60	20
3,000– 4,000	54	64	84	124	23	44	84	144	265	80	20
4,000– 5,000	63	73	93	133	24	48	93	158	288	100	25
5,000– 6,000	72	82	102	142	26	52	102	171	310	120	25
6,000– 7,000	81	91	111	151	28	57	111	184	331	140	30
7,000– 8,000	91	101	121	161	31	62	121	198	353	160	30
8,000– 8,500	103	113	133	173	34	68	133	215	380	180	35
8,500– 9,000	112	122	142	182	36	74	142	229	402	180	35
9,000– 9,500	127	137	157	197	40	81	157	250	435	200	35
9,500–10,000	152	162	182	222	43	88	172	272	471	200	35
10,000–10,500	190	210	230	270	48	96	192	300	515	210	35
10,500–11,000	—	225	245	285	—	106	212	329	562	225	35
11,000–11,500	—	255	275	315	—	119	237	365	621	255	35
11,500–12,000	—	290	310	350	—	131	262	401	679	290	35
12,000–12,500	—	330	350	390	—	144	287	436	735	330	35
12,500–13,000	—	375	395	435	—	156	312	471	789	375	35
13,000–13,500	—	425	445	485	—	169	337	506	843	425	35
13,500–14,000	—	480	500	540	—	181	362	543	905	480	35
14,000–14,500	—	540	560	600	—	200	400	600	1000	540	35

TABLE 2

Texas offshore allowables

Depth	*1967 depth yardstick b/d*		
	40 Ac.	*80 Ac.*	*160 Ac.*
0– 2,000	200	330	590
2,000– 3,000	220	360	640
3,000– 4,000	245	400	705
4,000– 5,000	275	445	785
5,000– 6,000	305	490	865
6,000– 7,000	340	545	950
7,000– 8,000	380	605	1050
8,000– 9,000	420	665	1150
9,000–10,000	465	730	1260
10,000–11,000	515	800	1380
11,000–12,000	565	875	1500
12,000–13,000	620	950	1625
13,000–14,000	675	1030	1750
14,000–15,000	735	1115	1880

Discovery Allowables:

1. Onshore—Extended from 18 months to 24 months. (Effective 3–28–66)
2. Onshore—Extended from 5 wells to 10 wells. (Effective 5–31–66)
3. Offshore discovery allowable is still limited to the 5 well and 18 month restriction.

Suggested minimum spacing for various units (Railroad Commission)

These figures are based on a rectangular shape with a length not more than twice the width and a well on the center line of adjacent units.

20 ac. = 330′ × 933′
40 ac. = 467′ × 933′ (Statewide = 467′ × 1200′)
80 ac. = 660′ × 1320′
160 ac. = 933′ × 1867′
320 ac. = 1320′ × 2640′
640 ac. = 1867′ × 3735′

Appendix B

Answer to Problem 1, Chapter 5

Successful prospect net cash flow

Year	*AFIT investment $M*	*Net income $M*	*Net cash flow $M*	*Cumulative net cash flow—$M*	
0	200	—	−200	−200	
1	50	—	−50	−250	
2	55	—	−55	−305	
3	195	—	−195	−500	
4	900	300	−600	−1,100	← Maximum neg. cash flow
5	100	500	+400	−700	
6		500	500	−200	
7		500	500	+300	← Cash flow becomes pos.
8		500	500	800	
9		500	500	1,300	
10		500	500	1,800	
11		500	500	2,300	
12		500	500	2,800	
13		300	300	3,100	
14		300	300	3,400	
15		100	100	3,500	

Maximum negative cash flow in year 4.
Cumulative cash flow becomes positive during year 7.
Payout in year 7.
Total after-tax investment: $M 1,500.
Actual value profit: $M 3,500.

Answer to Problem 1, Chapter 5

Investment cash flow stream—$M

	1	*2*	*3*	*4*	*5*	*6*	*7*		
Yr.	*Cap.*	*Tang.*	*Intang.*	*Exp.*	*Total col. 3 & 4*	*FIT Cr.*	*AFIT Inv.*	*Yr.*	*Remarks*
0	200						200	0	Lease bonus
							200		total, year 0
1	40			10	10	5	45	1	Seismic work
				10	10	5	5		lease rental
							50		total, year 1
2	50						50	2	Add. lease bonus
				10	10	5	5		lease rental
							55		total, year 2
3		90	210		210	105	195	3	Discovery well
							195		total, year 3
4		400	600		600	300	700	4	Dev. well
			200				200		lease facilities
							900		total, year 4
5				200	200	100	100	5	Dry hole
							100		total, year 5

(3) + (4) = (5) → (5) × .5 = (6) → (1) + (2) + (3) + (4) − (6) = 1500
AFIT investment

↑ Total AFIT investment for this prospect

Answer to problem 1, Chapter 6

10 doublings or roughly 73 years.

Answer to Problem 2, Chapter 6

	Cash flow streams—AFIT				*Discounting the cash flow* 15%		25%		30%	
	Investment	*Income*	*Net*	*Cumulative*	*Factors*	*Cash Flow*	*Factors*	*Cash Flow*	*Factors*	*Cash Flow*
0	−200		−200	−200	1.00	−200	1.00	−200	1.00	−200
1	−900	200	−700	−900	0.90	−630	0.80	−560	0.75	−525
2	−100	300	200	−700	0.75	150	0.65	130	0.60	120
3		400	400	−300 *	0.65	260	0.50	200	0.45	180
4		400	400	+100	0.55	220	0.40	160	0.35	140
5		300	300	400	0.50	150	0.33	100	0.25	75
6		300	300	700	0.45	135	0.30	90	0.20	60
7		200	200	900	0.40	80	0.24	48	0.16	32
8		100	100	1,000	0.35	35	0.18	18	0.12	12
9		100	100	1,100 ↑ Actual Value Profit	0.30	30	0.14	14	0.09	9
Totals	−1,200	2,300	1,100			+230		−0−		−97

* Payout = 3.75 years
Maximum negative cash flow = −900
DCFR = 25%

Answer to Problem 1, Chapter 9

$—0.05 per roll.

Answer to Problems 1, 2, 3—Chapter 10

Sensitivity checks
School prospect
Thousands of dollars

	Base case			*Case 2*		
	Cash flow streams—AFIT			*Two × exploration expenditures*		
				Net cash	*Discounted @*	
	Investment [a]	*Income*	*Net*	*Flow—AFIT*	*9%* [b]	*10%*
0	−500		−500	−1,000	−1,000	−1,000
1	−200		−200	−400	−367	−364
2	−600		−600	−1,200	−1,010	−991
3	−600		−600	−1,200	−926	−901
4	−1,400	1,200	−200	−200	−142	−137
5	−500	1,200	700	700	455	435
6		1,200	1,200	1,200	715	677
7		1,200	1,200	1,200	656	616
8		1,200	1,200	1,200	602	560
9		600	600	600	276	254
10		600	600	600	253	232
11		600	600	600	232	210
12		600	600	600	214	191
13		600	600	600	196	174
Totals	−3,800	9,000	5,200	3,300	+154	−44
	DCFR—22.2%			DCFR—9.8%		

	Case 3				*Case 4*			
	Two × development expenditures				*Two-year delay of revenue*			
	Net Cash	*Discounted @*			*Net Cash*	*Discounted @*		
	Flow—AFIT	*10%*	*13%*	*14%*	*Flow—AFIT*	*10%*	*14%*	*15%*
0	−500	−500	−500	−500	−500	−500	−500	−500
1	−200	−182	−177	−175	−200	−182	−175	−173
2	−600	−496	−470	−461	−600	−496	−461	−454
3	−600	−451	−416	−404	−600	−451	−404	−394
4	−1,600	−1,093	−981	−947	−1,400	−956	−829	−799
5	200	124	109	104	−500	−311	−260	−249
6	1,200	677	576	547	1,200	677	547	518
7	1,200	616	510	480	1,200	616	480	450
8	1,200	560	451	421	1,200	559	421	391
9	600	254	199	184	1,200	509	369	341
10	600	232	176	162	1,200	463	324	296
11	600	210	156	142	600	210	142	128
12	600	191	138	124	600	191	125	112
13	600	174	124	109	600	174	109	97
14					600	158	96	85
15					600	143	84	73
Totals	3,300	316	−105	−214	5,200	804	68	−78
	DCFR [c]—12.1%				DCFR—14.5%			

[a] Years 0–3 = exploratory expenditures.
[b] Check the desk calculator shortcut for discounting a cash flow, Appendix B.
[c] Three point plot, Figure 23.

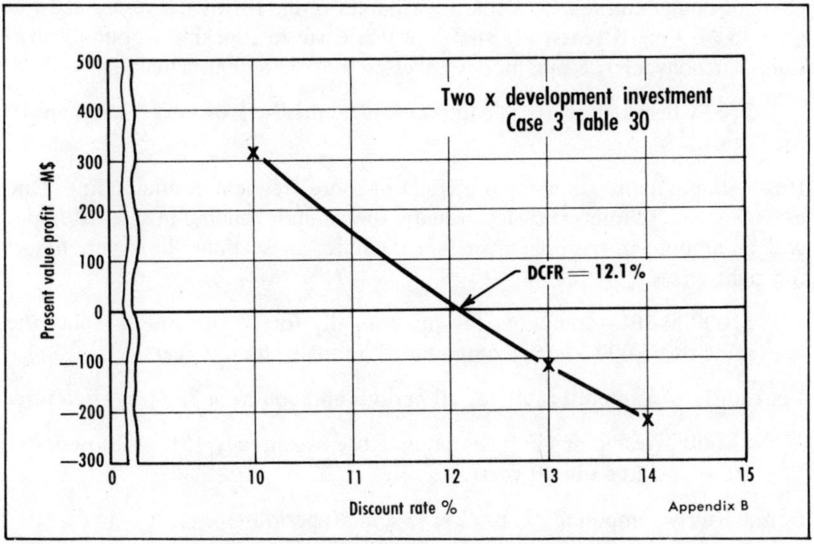

FIG. 30 GRAPHIC SOLUTION FOR DCFR
TWO×DEVELOPMENT INVESTMENT

Appendix C

In chapter 6, p. 74, we spoke of the effect of variations in compounding frequency. At that time the relative effect of various compounding frequencies was not shown. A simple illustration follows which will demonstrate the effect of compounding. On Table 19, Chapter 6, the FV of 1 invested at 10% grew to 2.59 in 10 years. We shall use this basic relationship in our illustration. For convenience our money invested will be $1000. Then:

> $1000 invested at 10% compounded annually grows to $2,594 in 10 years.

It is a simple matter to show the effect of more frequent compounding. One can use standard interest tables. Usually the column heading in interest tables will be amount of 1. You must look in different sections, however, to get the right answer.

> $1000 at 10% compounded semi-annually for 10 years is in reality the same as $1,000 at 5% compounded annually for 20 years.

Look at the 5% interest table for 20 periods and you have $2,653. Therefore,

> $1,000 invested at 10% compounded semi-annually (5%—20 periods) grows to $2,653 in 10 years.

For quarterly compounding, use 2½% and 40 periods:

> The answer is $2,685.

For monthly compounding (0.8⅓% and 120 periods):

> $1000 at 10% for 10 years grows to $2,707.

Let's check the effect of more frequent compounding. The net effect of even monthly compounding is $113 in 10 years. The additional interest comes in over the entire ten years with most in the latter years. The $113 is 4% more than the total money after annual compounding.

As discussed in Chapters 5 and 8, the dominance of several key parameters relegates small differences (such as income from more frequent compounding) to insignificance.

Appendix D

One of our goals was to illustrate without the computer. Since a Monte Carlo simulation requires a computer, this section is relegated to an appendix. As was the case in the two chapters on risk analysis, it draws from many sources, but in particular from unpublished work of G. Rogge Marsh.[1] The basic concepts involving Monte Carlo simulation are not new and can be found in most recent textbooks on statistics.

The term "Monte Carlo analysis" was spawned by two mathematicians, John von Neumann and S. Ulam. They developed the technique to solve a problem on nuclear shielding which was too complex for analytical solution and too risky for empirical solution.[2]

What Monte Carlo simulation does In Monte Carlo simulation, a specific operation is mathematically performed thousands of times.

Input data involve estimates of the range of probabilities for the variables. By means of a random number generator, specific values for each variable are chosen at frequencies described by input data. Values for each variable are generated and the resultant answer to the mathematical operations contains the complete range of probable solutions. The answer takes the form of a cumulative probability distribution curve.

We will now backtrack a bit and explain more thoroughly some of the terms.

I. Random numbers

First let's explain a random number generator and why we want to use one. Remember our 27 case tree diagram, Fig. 19? It showed 27 possible combinations of three values for three parameters; but there are really more than 27 possibilities. What is more, we may not have made the best combination of the various values. If not, would 1,000 cases give a different or better answer? Whenever we have problems involving many combinations of variables Monte Carlo simulation is useful. It will show us the most probable answer based on a sufficient quantity of cases to test the limits of all variables.

Most computers have a subroutine which will generate random numbers. (You could think of the last two digits in a column of telephone numbers as a random number generator.) Numbers are selected between 0 and 1.0 and related to the input data which is submitted in the form of a frequency distribution. The computer converts the input data, a frequency distribution, into a cumulative probability distribution. The numbers chosen then relate to the cumulative probability; thus, although the number is randomly chosen, the frequency of occurrence represents a judgment by the geologist. The resultant values *do* reflect his estimate of the most probable occurrences.

II. The input

The geologists' judgments enter the computer by means of frequency distributions. Remember, a frequency distribution is easily converted to a probability distribution. Often these distributions are no more than triangular distribution of the minimum, maximum, and most likely.

Fig. 31 illustrates the probability distributions used for the School project. The three variables for each parameter, representing our minimum, maximum, and most likely, are shown by circles. Note carefully the following:

1. The three values used previously no longer represent the lowest or highest possible values. For example in recovery factors we assigned probabilities to values of 0.7, 1.0 and 1.5 $\overline{M}$cf/acre-ft. When these three points are placed in triangular distribution the minimum point becomes 0.4 $\overline{M}$cf/acre-ft. and the maximum point becomes 1.6 $\overline{M}$cf acre-ft. These points, on the X axis, are absolute. No value can be higher than 1.6 $\overline{M}$cf or lower than 0.4 $\overline{M}$cf/acre-ft.

2. Does the diagram on recovery factor help you see that there are, within this triangular distribution, an infinite number of possibilities between 0.4 and 1.6 $\overline{M}$cf/acre-ft.?

3. Knowing that the computer converts these triangular distributions to cumulative probability distributions, does it also show you that—despite the many possible values that the geologists' estimates of probability are in control? No matter how many values are chosen by the computer, the probabilities established by the explorationist are in command. The randomly selected cases will still show the highest probability around 1.0 $\overline{M}$cf/acre-ft.

4. The distributions do not have to be triangular. If triangular they do not have to be isosceles triangles, i.e., the most likely case does not have to be exactly halfway between the minimum and maximum. A rectangular distribution means you think there is no most likely value and that the minimum and maximum values have the same probability of occurrence. Distributions can be continuous—curved rather than triangular. They can be discrete, in which case the variables can have only certain distinct values. There is nothing magic about triangular distributions. They just result from three values and many explorationists think more easily in three-value terms. See Fig. 32 for distribution examples.

5. Note the net pay diagram on Fig. 31. The three probabilities used in Fig. 20 cause line a to intersect the Y axis. Mathematically this says we expect some negative values, an impossibility. (The minimum becomes −25 ft.) The moral is this—if you plan to submit triangular distributions (based

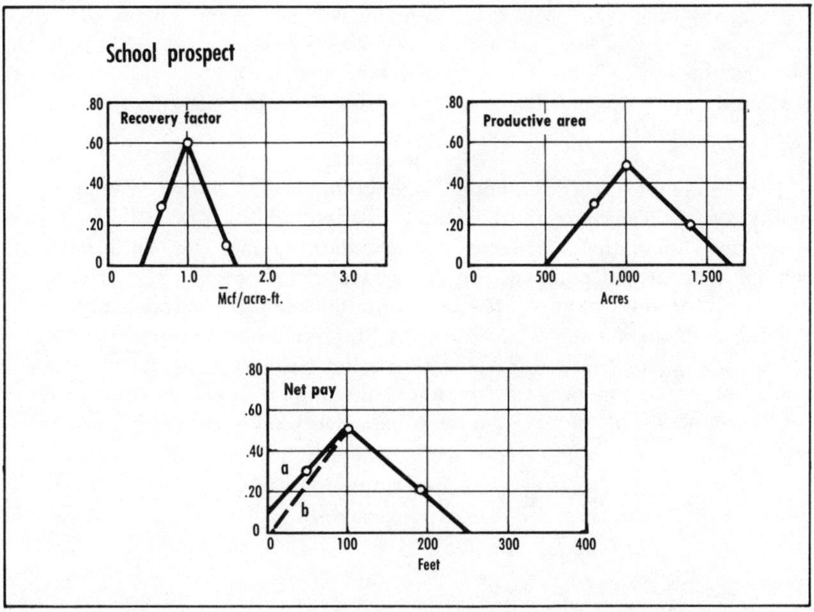

FIG. 31 PROBABILITY DISTRIBUTIONS FOR FIELD SIZE

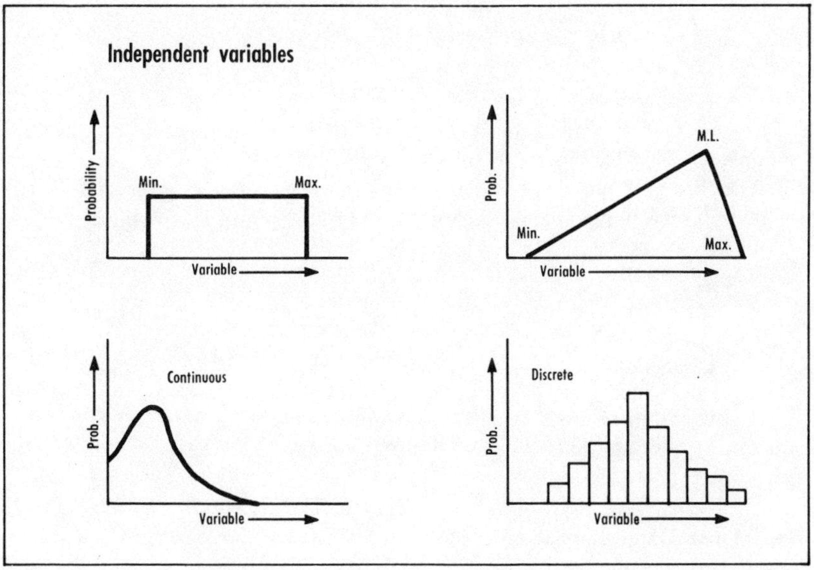

FIG. 32 TYPES OF DISTRIBUTIONS

on probabilities other than maximum and minimum) to a computer, plot them first to make sure you are not assigning impossible values. Line b can be substituted with little change in end results. It is close to but not zero at the X axis and raises the 30% probability from 50 ft. to 60 ft.

III. The output

The computer takes a random number for a variable, reads its probability of occurrence from the cumulative probability distribution created from data submitted. It reads corresponding values for each variable, multiplies them and computes one value for the answer desired. It then repeats this process many times—each time choosing a random number but relating it to the geologists' judgment as to what values were most probable.

Our triangular distributions for the School prospect were entered in a computer simulation program. In about 20 seconds 5,000 reserve size cases were computed. The output from the program was as follows:

Reserve-BCF	*Probability*
29.8	.261
50.4	.232
68.1	.203
83.9	.174
100.0	.145
117.7	.116
138.8	.087
168.1	.058
208.0	.029
410.1	.001

The results are plotted on Fig. 33 in the same manner as our 27 case example of Fig. 22. Note the smoothness from 5,000 cases compared to the 27. The distributions are similar however. The average reserve and risk weighted reserve are compared below for the two types of calculations.

	Hand calculation 27 cases	*Computer run 5,000 cases*
Average reserve	129	111
Risk weighted reserve	29.2	32.0

The output from Monte Carlo simulation is in the form of a cumulative probability distribution for the desired answer. The case illustrated was reserve size.

Why Monte Carlo simulation There are benefits from Monte Carlo simulation:

1. The frequency of occurrence of all critical variables can be handled

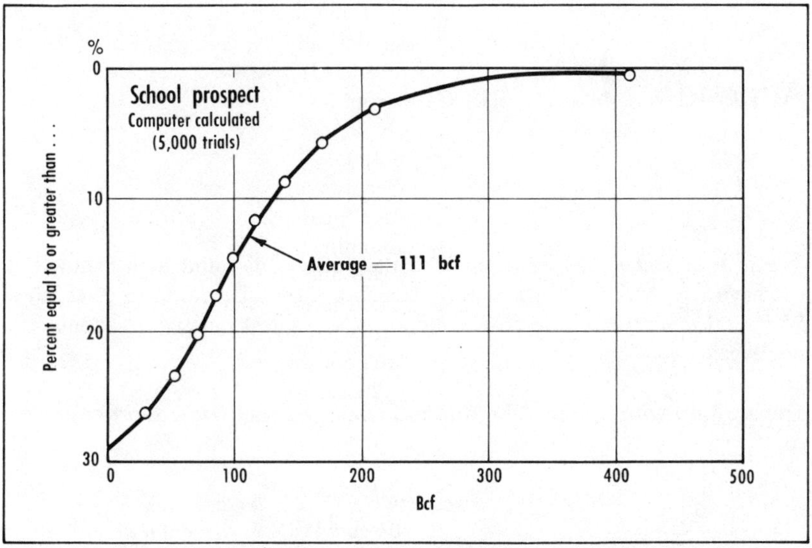

FIG. 33 CUMULATIVE PROBABILITY DISTRIBUTION

simultaneously and randomly based on input judgment. The answer received is superior to a sensitivity check where only one variable can be tested at a time; however, you still may wish to test the sensitivity of one or more key variables separately.

2. The range of possible values derived from input probability distributions shows that you really have no such thing as a single most likely case. The steepest slope of the cumulative probability distribution is the most likely range of values. However, there are instances where the slope is never very steep. In such cases the range of yardsticks (DCF, PVP, etc.) may be broad at the higher frequencies. If this is true, you need to know this characteristic of your investment opportunity.

3. You will better describe an opportunity, learning more about the high side potential and the low side risks.

If you have a computer available, you may wish to explore this technique. It has interesting possibilities for those willing to take the time to understand its strengths and weaknesses.

BIBLIOGRAPHY

1. Marsh, G. Rogge, "Risk Assessment Terminology," 1970 unpublished paper.
2. Springer, Clifford H. et al, "Probabilistic Models," Vol. 4 of Mathematics for Management Series 1968, Richard D. Irwin, Inc., p. 135.

Appendix E

You may sometimes be faced with the need to discount a cash flow but not have any discount tables handy. Here's a shortcut method to establish a PVP by discounting a cash flow without tables—a desk calculator is required, however.

The method with tables For our basic data, we shall use a short cash flow for 5 years.

Discounting with 10% factors

Year	*Net cash flow—$\overline{M}$\$*	*Discount factors 10%*	*Discounted cash flow—$\overline{M}$ \$ @ 10%*
1	100	.909	91
2	100	.836	83
3	100	.751	75
4	100	.683	68
5	100	.621	62
	500		379

With the 10% discount factors, the present value of the income stream is shown to be \$379 million.

The shortcut—sans tables The shortcut method involves only the first year discount factor—which is $\frac{1}{(1+i)^1}$ or $\frac{1}{(1+i)}$. In the case of 10% tables, this number is $\frac{1}{1+0.1} = \frac{1}{1.1} = 0.909$. To apply the shortcut, start at the year of the final cash flow (year 5 in our case) and calculate as follows: multiply income in last year times the first year factor, add the next earlier year's income, multiply by same factor, etc. until last year (1) has been added and multiplied. A sample calculation is shown below.

Shortcut discounting method

Year	Net cash flow—$\overline{M}$\$		Order of calculation
1	100	100 + 317 = 417 × .909 = 379	Step 5
2	100	100 + 249 = 349 × .909 = 317	Step 4
3	100	100 + 174 = 274 × .909 = 249	Step 3
4	100	100 + 91 = 191 × .909 = 174	Step 2
5	100	100 × .909 = 91	Step 1

The result is exactly the same as having all of the discount factors—$379 million. Leave the proof to the mathematicians. The method works.

Index